Advances in Information Security

Volume 71

More information about this series at http://www.springer.com/series/5576

Anne V. D. M. Kayem • Stephen D. Wolthusen
Christoph Meinel
Editors

Smart Micro-Grid Systems Security and Privacy

Editors
Anne V. D. M. Kayem
Hasso-Plattner-Institute, Faculty of Digital Engineering
University of Potsdam
Potsdam, Germany

Christoph Meinel
Hasso-Plattner-Institute, Faculty of Digital Engineering
University of Potsdam
Potsdam, Germany

Stephen D. Wolthusen
Department of Mathematics and Information Security
Royal Holloway, University of London
Egham, Surrey, UK

Norwegian Information Security Laboratory
Gjovik University College,
Norwegian University of Science and Technology
Trondheim, Norway

ISSN 1568-2633
Advances in Information Security
ISBN 978-3-030-08252-9 ISBN 978-3-319-91427-5 (eBook)
https://doi.org/10.1007/978-3-319-91427-5

This Springer imprint is published by the registered company Springer Nature Switzerland AG
The registered company address is: Gewerbestrasse 11, 6330 Cham, Switzerland

. . . To Scientists in the field of Cyber-Physical Systems. . .

Preface

People who work in the field of energy management have the pleasure of working on a topic whose results are visualisable and beneficial to society. There is also the payoff of knowing that economic growth, and perhaps even life as we know it, would be impossible without power or energy. Energy grid models play a crucial role in idealisations of real-world economies.

Until recently, however, energy grid management systems have largely remained the preserve of electrical engineers and researchers. Its concepts are not really esoteric or difficult, but they are relatively new to the computer science community, so it has taken a while to sort out the best ways of designing energy grids that can be controlled using a cyber system.

Now after more than 30 years of development, smart grids and micro-grid technologies have matured to the point where they are ready to take their place in discussions on computing that are centred on matters of security and privacy. This book is intended to provide an overview of some of the primary techniques that can be used to model adversarial scenarios in both smart grids and micro-grids. In designing energy systems to operate as a combination of a cyber (algorithms and software systems to control energy generation hardware) and physical system (energy generation components and grid), we now find ourselves having to handle aspects such as data manipulations to enable energy theft, masking adversarial behaviours as faulty behaviour, price signal manipulation, and inference of private user behaviours, to name but a few potential security and privacy vulnerabilities. The material covered, in terms of adversarial scenarios, draws from classical attacks centred on energy theft, misattribution, and grid destabilisation. The focus is on how these attacks, masked as system failures or component malfunctions, can be used to cause the breakdown of energy grids without drawing attention to the adversary.

We assume that the reader has some familiarity with basic concepts in computer science, security and privacy, and smart grid technologies. In a nutshell, the reader should be able to write programs and have some understanding of energy flow control manipulation. Otherwise, the book is intended to be self-contained.

This book is meant to be used as a reference manual for researchers and students, in need of concrete examples on how to model malicious scenarios in smart girds and micro-grids. The book can also be used to introduce graduate students to the field of security and privacy in smart grids and micro-grids. Supplemented by papers from the literature, the book can also serve as the basis for an introductory graduate course on cyber-physical systems, or as the basis for self-study by researchers in the fields of cyber-physical systems, resource constrained computing, and smart grids/micro-grids, who want access to the literature in this field.

Related Books Related texts include *Smart Grid Infrastructure and Networking by Iniewski; Distributed Algorithms by Lynch; Introduction to Algorithms by Cormen, Leiserson, Rivest, and Stein; and Fault Tolerance in Distributed Systems by Jalote.* This book could be considered as supplementary to each of these in studying smart grids/micro-grids particularly ones in which management is distributed.

How to Use This Book Since readers of this book are likely to come from different backgrounds, being aware of the implicit structure of this book might be helpful. With this in mind, Chap. 1 puts the material of the book into perspective and will help readers understand the basic objectives of the book as well as the role of the remaining chapters in meeting those objectives. Chapters 2 and 3 are focused on presenting attacks and countermeasures on state estimation, as well as an example of an authentication protocol in smart grids. Chapter 4 presents a survey of potential vulnerabilities in authentication protocols for smart grids highlighting the similarities with standard authentication systems. Chapters 5–7 discuss micro-grid architectures, focusing on the special case of resource constrained smart micro-grids. Resource constrained smart micro-grids are a special case of micro-grids designed to operate autonomously in rural/remote environments where connectivity to standard smart micro-grids is logistically or economically infeasible. Since such micro-grids are typically supported by a lossy communications network, adversarial scenarios must be modelled to account for unreliability, and special properties of flow control identified in order to differentiate benign faulty behaviours from malicious attempts at subversion.

We hope that you will find this book rewarding in many ways, and that it will serve as a basis for even more exciting discoveries on this topic.

Potsdam, Germany — Anne V. D. M. Kayem
Potsdam, Germany — Christoph Meinel
London, UK — Stephen D. Wolthusen
March 2018

Acknowledgements

We begin by expressing our heartfelt gratitude to the Norwegian Research Council, South African National Research Foundation, and the Hasso-Plattner-Institute for the funding and infrastructural support provided to facilitate this work. This book would not have been possible without your support!

In addition, we would like to thank all the reviewers, for taking the trouble to painstakingly provide feedback to the authors both on the chapter proposals and the full chapter submissions. In particular, special thanks to Andrew Adamatzky (University of West England, UK), Chris Mitchell (Royal Holloway, University of London, UK), Cristina Alcaraz (University of Malaga, Spain), Dieter Hutter (University of Bremen, and DFKI, Germany), Ingo Stengel (University of Applied Sciences, Karlsruhe, Germany), Martin Strohmeier (University of Oxford, UK), Stephen Marsh (University of Ontario Institute of Technology, Canada), Sule Yildirim-Yayilgan (Norwegian University of Science and Technology, Norway), Sylvia Osborn (University of Western Ontario, Canada), Trupil Limbasiya (Birla Institute of Technology & Science (BITS), India), and Zeyar Aung (Masdar Institute, Khalifa University of Science and Technology, UAE).

Compiling a contributed volume requires not only content from the authors but also a commitment to deliver high-quality material in a timely manner. We would like to take this opportunity to express our heartfelt gratitude to all the authors for making this a pain-free process. Special thanks to Ammara Gul (Royal Holloway, University of London, UK); Anesu Marufu and Pacome Ambassa (University of Cape Town, South Africa); Stephen Wolthusen (Royal Holloway, University of London, UK, & Norwegian University of Science and Technology, Norway); Trupil Limbasiya, Aakriti Arya, and Pragya Verma (NIIT University, India); and Hikaru Kishimoto (Osaka University, Japan), Naoto Yanai (Osaka University, Japan), and Shingo Okamura (National Institute of Technology, Nara College, Japan). Thank you for your patience and contributions.

While we were compiling this book, we received informal feedback from several colleagues. In particular, we would like to thank Leonard Barolli, Sylvia Osborn, Dieter Hutter, and Tei-Wei Kuo for their constructive feedback and suggestions.

Finally, we would also like to thank our editorial managers Susan Lagerstrom-Fife, Jennifer Malat, and Caroline Flanagan for providing the editorial support needed to compile this book. You seemed to know exactly when to keep quiet to let us get on with our work, and when to push for deliverables!

Contents

Contributors

Pacome L. Ambassa Department of Computer Science, University of Cape Town, Rondebosch, Cape Town, South Africa

Aakriti Arya NIIT University, Neemrana, Rajasthan, India

Ammara Gul Department of Mathematics and Information Security, Royal Holloway University of London, Egham, Surrey, UK

Anne V. D. M. Kayem Hasso-Plattner-Institute, Faculty of Digital Engineering, University of Potsdam, Potsdam, Germany

Hikaru Kishimoto Osaka University, Suita, Osaka, Japan

Trupil Limbasiya Birla Institute of Technology & Science (BITS), Pilani, Goa, India

Anesu M. C. Marufu Department of Computer Science, University of Cape Town, Cape Town, South Africa

Christoph Meinel Hasso-Plattner-Institute, Faculty of Digital Engineering, University of Potsdam, Potsdam, Germany

Shingo Okamura National Institute of Technology, Nara College, Yamatokoriyama, Nara, Japan

Stephen D. Wolthusen Department of Mathematics and Information Security, Royal Holloway, University of London, Egham, Surrey, UK

Norwegian Information Security Laboratory, Gjovik University College, Norwegian University of Science and Technology, Trondheim, Norway

Naoto Yanai Osaka University, Suita, Osaka, Japan

List of Reviewers

Andrew Adamatzky University of West England, UK

Anne V. D. M. Kayem Hasso-Plattner-Institute, Germany

Chris Mitchell Royal Holloway, University of London, UK

Cristina Alcaraz University of Malaga, Spain

Dieter Hutter University of Bremen, and DFKI, Germany

Ingo Stengel University of Applied Sciences, Karlsruhe, Germany

Martin Strohmeier University of Oxford, UK

Stephen Marsh University of Ontario Institute of Technology, Canada

Sule Yildirim-Yayilgan Norwegian University of Science and Technology, Norway

Sylvia Osborn University of Western Ontario, Canada

Trupil Limbasiya Birla Institute of Technology & Science (BITS), India

Zeyar Aung Masdar Institute, Khalifa University of Science and Technology, UAE

Chapter 1
Power Systems: A Matter of Security and Privacy

Anne V. D. M. Kayem, Stephen D. Wolthusen, and Christoph Meinel

Abstract Studies indicate that reliable access to power is an important enabler for economic growth. To this end, modern energy management systems have seen a shift from reliance on time-consuming manual procedures, to highly automated management, with current energy provisioning systems being run as cyber-physical systems. Operating energy grids as a cyber-physical system offers the advantage of increased reliability and dependability, but also raises issues of security and privacy. In this chapter, we provide an overview of the contents of this book showing the interrelation between the topics of the chapters in terms of smart energy provisioning. We begin by discussing the concept of smart-grids in general, proceeding to narrow our focus to smart micro-grids in particular. Lossy networks also provide an interesting framework for enabling the implementation of smart micro-grids in remote/rural areas, where deploying standard smart grids is economically and structurally infeasible. To this end, we consider an architectural design for a smart micro-grid suited to low-processing capable devices. We model malicious behaviour, and propose mitigation measures based properties to distinguish normal from malicious behaviour.

Keywords Lossy networks · Low-processing capable devices · Smart micro-grids · Security · Privacy · Energy

A. V. D. M. Kayem (✉) · C. Meinel
Hasso-Plattner-Institute, Faculty of Digital Engineering, University of Potsdam, Potsdam, Germany
e-mail: christoph.meinel@hpi.uni-potsdam.de

S. D. Wolthusen
Department of Mathematics and Information Security, Royal Holloway, University of London, Egham, Surrey, UK

Norwegian Information Security Laboratory, Gjovik University College, Norwegian University of Science and Technology, Trondheim, Norway
e-mail: stephen.wolthusen@rhul.ac.uk

A. V. D. M. Kayem et al. (eds.), *Smart Micro-Grid Systems Security and Privacy*, Advances in Information Security 71, https://doi.org/10.1007/978-3-319-91427-5_1

1.1 Context and Motivation

Smart grids offer a cost-effective approach to fair and equitable power provisioning in urban areas [10, 11]. However, deploying smart grids in rural and remote areas that have no and/or intermittent access to national power networks, can be both expensive and logistically challenging. Proposed solution measures recommend using smart micro-grids based on distributed renewable energy sources, instead, as a suitable and economically feasible alternative [16, 17]. Since the energy sources for such micro-grids are variable and only partly predictable, using a combination of energy management techniques is important in maintaining grid stability. However, in contrast to smart grids, the inherently distributed architecture of the energy resources on such micro-grids, implies the absence of a centralised trusted grid monitoring and management point. As such, grid management is handled via a distributed system which is vulnerable to grid subversion attacks geared primarily at energy theft. A further concern is that generation and feed-in mis-recordings can be exploited to reveal sensitive information which in turn can be used to provoke privacy violations. Protecting against energy theft [16, 17] and privacy violation attacks [5] is important in guaranteeing grid usability, trust, and reliability which are needed to ensure grid stability.

This chapter we provide an overview of the contents of this book. We begin by discussing the concept of smart-grids in general, proceeding to narrow our focus to smart micro-grids in particular. The discussion on smart grids is deepened in Chaps. 2–4, with discussions on state estimation, attacks on state estimation, and authentication. In Chaps. 5 and 6, lossy networks are considered as an enabler for the implementation of smart micro-grids in remote/rural areas. However, the unreliable nature of the network raises new attacks models that must be studied to protect against energy theft, and privacy violations. Potential attack models are therefore studied, in the context of smart micro-grids that are run via lossy networks, and mitigating measures proposed. In the next section, we provide a brief overview smart grids in general and smart micro-grids briefly highlighting the advantages and challenges of both from the energy management perspective.

1.2 Smart Grids and Smart Micro-Grids

Several factors have resulted in the shift from a resource controlled, to a technology controlled model in smart-grids and micro-grids [7]. A key driving force is the notion of sustainability and fairness in power distribution. Furthermore, the current transformation has the advantage of enabling environmentally friendly ways of managing energy resources. Smart grids have in some sense enabled this, and increasingly in smart micro-grids. A smart micro-grid is a small version of a smart grid, that is customised to match the energy demands of a specific environment using intelligent controls, optimisation solutions, generation resource management, as well as power marketing, for instance. In a smart micro-grid, the idea is to control a group of interconnected loads and distributed energy resources within clearly defined electrical boundaries as a single controllable entity with respect to the grid. Generally speaking the micro-grid should be able to operate both in grid connected and island mode.

One prerequisite for grid stability is the balance between energy consumption and generation. However, what is often neglected is the element of security and perhaps when one considers the pervasive computing environment in which smart grids and micro-grids operate, privacy is also an important aspect to take into account in ensuring grid stability. Furthermore, energy generation flows no longer follow the time variant consumption pattern where energy flow was typically uni-directional from the generation source to the consumer. In smart energy platforms, energy may flow from the consumer (e.g. one owning a renewable energy generation facility) into the grid, and vice versa. This has resulted in a complex grid structure, which is further challenged by having to handle intermittent and unpredictable energy supply sources. Other considerations include upgrading old electrical infrastructure to cope with the technological requirements of the platforms.

While both smart grids and micro-grids, both provide a means of savings in terms of energy costs, designing micro-grids that are able to operate independently over long periods requires careful consideration. Furthermore, as discussed in the next section and throughout this book, taking into account the security and privacy considerations is important. We briefly discuss some of the security and privacy considerations in the following section.

1.3 State Estimation and Authentication in Smart Grids

Problems of security and privacy in smart grids and micro-grids can be categorised into three groups. In the first, one sees adversarial manipulations that are aimed primarily at data distortion. This appears in consumption data manipulation, and pricing data manipulation, for instance [2, 3, 6]. The second category involves, dealing with impersonation where malicious users aim to shift power consumption to benign users. the goal here is to obtain more than the allocated quota of power, and/or avoid paying for power consumed. Finally, in the third category we see attacks aimed at studying private user behaviours in order to design attacks centred primarily around impersonation. In the following subsections, we briefly consider the types of attacks discussed in this book and how these impact on smart grids and micro-grids in general.

1.3.1 Attacks on Power State Estimation

In discussing the issue of attacks on power state estimation, Gul and Wolthusen (see Chap. 2) seek to provide an integrated, up-to-date survey of various attack models and corresponding protection against state estimation in larger-scale power systems [1, 4, 9]. As mentioned before, electrical power grids supported by smart technologies are the de-facto standard for providing energy. However, in introducing computing technologies into the infrastructure of electrical power systems there is also the possibility of manipulating such systems adversarially. Gul and Wolthusen, begin by providing a brief overview of various attack models drawn from the literature on smart grids. In particular, they compared the models and the mitigating solutions proposed to make apparent the ones that are most relevant and practically implementable. The comparisons are performed on account of complexity level, optimality, considered structure, topology knowledge requirements and practical implementation along with the impact of advanced metering devices.

1.3.2 Authentication in Smart Grids

Following on the Gul and Wolthusen discussion on attacks on power state estimation, Naoto et al. (see Chap. 3) discuss the smart grid as a method of allowing users access information related to their electricity usage via IP networks. In this case, both the validity and the privacy of such information needs to be guaranteed in order to ensure grid stability and user participation. The idea is that this model can be used to charge consumers' electricity bills directly to them via the Smart Grid, even when the users are outside their homes. Billing information is tightly intertwined with consumer privacy; hence, in Chap. 3, Naoto et al. propose an anonymous authentication protocol for electricity usage on the Smart Grid. Their main idea is to utilise group signatures with controllable linkability. In these group signatures, only designated signers can generate digital signatures with anonymity under a single group public key, and only entities with a link key can distinguish whether the signatures are generated by the same signer or not. Whereas their proposed protocol can include any group signature scheme with controllable linkability, Naoto et al. also propose new controllably linkable group signatures with tokens, which are handled by smart meters on the Smart Grid. Naoto et al. implement the proposed group signatures, and provide an estimate of the computational time of their anonymous authentication protocol on Raspberry Pi.

1.3.3 Attacks on Authentication in Smart Grids

Limbasiya et al. (see Chap. 4) wrap up the contributions of this book by providing an overview of the evolution of a conventional electric grid infrastructure which can be dated back to 1880s when the outstanding sources of energy were hydraulics and gas energy. They concur with the general literature on smart grids and smart micro-grids, where the general perception is that one cannot only depend upon the classic electric grid system in today's digital world. Using smart technologies to control electrical grids offers advantages such as enabling active participation in energy management by consumers, accommodating new energy generation facilities, and anticipating system disturbances, for instance. The smart grid and smart micro grid provide electric power in many efficient and measured ways, which is important in the technology-enabled market. As such, Limbasiya et al. explain the structure of the smart grid and discuss the various authentication schemes associated with it, as well as describing different security parameters and varied attacks, which should be considered for successful and secure use of smart grids and micro-grids.

1.4 Resource Constrained Smart Micro-grids

Resource constrained smart micro-grids describe a subset of smart micro-grid architectures in which communications are handled over a lossy network. As mentioned before, the primary reason for doing this is to design an economically and structurally feasible powering alternative, suited to regions where connectivity to standard smart grids is impaired by factors such as load-shedding. Chaps. 5–7 discuss issues such as power flow control, marketing, energy theft, and privacy. We discuss the interrelations between each in the following sections.

1.4.1 Architectures Matter

Following on Naoto et al.'s work, Kayem et al. (see Chap. 5) present a resource constrained smart micro-grid architecture to describe a class of smart micro-grid architectures that handle communications operations over a lossy network and depend on a distributed collection of power generation and storage units [8]. Disadvantaged communities with no or intermittent access to national power networks can benefit from such a micro-grid model by using low cost communication devices to coordinate the power generation, consumption, and storage [15]. Furthermore, this solution is both cost-effective and environmentally-friendly. One model for such micro-grids, is for users to agree to coordinate a power sharing scheme in which individual generator owners sell excess unused power to users wanting access to power. Since the micro-grid relies on distributed renewable energy generation sources which are variable and only partly predictable, coordinating micro-grid operations with distributed algorithms is necessity for grid stability. Grid stability is crucial in retaining user trust in the dependability of the micro-grid, and user participation in the power sharing scheme, because user withdrawals can cause the grid to breakdown which is undesirable. In this chapter, we present a distributed architecture for fair power distribution and billing on micro-grids. The architecture is designed to operate efficiently over a lossy communication network, which is an advantage for disadvantaged communities. We build on the architecture to discuss grid coordination notably how tasks such as metering, power resource allocation, forecasting, and scheduling can be handled. All four tasks are managed by a feedback control loop that monitors the performance and behaviour of the micro-grid, and based on historical data makes decisions to ensure the smooth operation of the grid [18]. Finally, since lossy networks are undependable, differentiating system failures from adversarial manipulations is an important consideration for grid stability. To this end, Kayem et al. provide a characterisation of potential adversarial models and discuss possible mitigation measures.

1.4.2 Power Auctioning and Cheating

Anesu et al. (see Chap. 6), provide a framework to specify how cheating attacks can be conducted successfully on power marketing schemes [12–14, 19] in resource constrained smart micro-grids based on a model such as the one proposed by Kayem et al. (see Chap. 4). This is an important problem because such cheating attacks can destabilise and in the worst case result in a breakdown of the micro-grid. Anesu et al. consider three aspects, in relation to modelling cheating attacks on power auctioning schemes. First, they aim to specify exactly how in spite of the resource constrained character of the micro-grid, cheating can be conducted successfully. Second, they consider how mitigations can be modelled to prevent cheating, and third, they discuss methods of maintaining grid stability and reliability even in the presence of cheating attacks. Moreover, Anesu et al. use an Automated-Cheating-Attack (ACA) conception to build a taxonomy of cheating attacks based on the idea of adversarial acquisition of surplus energy. Adversarial acquisitions of surplus energy allow malicious users to pay less for access to more power than the quota allowed for the price paid. The impact on honest users, is the lack of an adequate supply of energy to meet power demand requests. They conclude with a discussion of the performance overhead of provoking, detecting, and mitigating such attacks efficiently.

1.4.3 Inferring Private Behaviours

In Chap. 7, Ambassa et al. conclude this book with an overview of mechanisms to infer private user behaviours on resource constrained smart micro-grids. As mentioned before, resource constrained smart micro-grid architectures handle communications over distributed lossy networks to minimize operation costs. However, the unreliable nature of lossy networks makes inferring private user behaviours comparatively easier than in standard smart grids and micro-grids, to infer private user behaviours based on leaked information. Applying existing data perturbation anonymisation approaches that work by distorting the data with additive noise makes distinguishing deliberate noise additions from system, and malicious additive noise, a challenging problem. Ambassa et al. present a brief survey of how privacy inferences can be drawn, and propose a mitigation method.

1.5 Discussions

The field of Smart grids and micro-grids have seen rapid strides over the past decade, particularly from the perspective of computing science. Moreover, the idea of the "Internet-of-Things" has and continues to play a considerable role in the design of the networking architectures that underpin data transmissions on these systems. While secure and privacy preserving data transmission has been studied extensively in the conventional fields of Networking and Data Security, for specialists in the field of cyber-physical systems issues such as distribution, scalability, reliability, usability and availability, for instance, are extremely important. The emergence of cyber-physical systems as a discipline requires re-thinking standard computing solutions including ones that are provably undecidable and/or for which no optimal polynomial-time solution exists. In typical applications, the types of attacks provoked can range from the fairly simple, to the inordinately complex. For example, provoking a denial-of-service attack in a sensor controlled smart micro-grid, might simply require creating a scenario in which the sensor batteries are depleted. On the more sophisticated end one could envisage attacks that are masked as system faults and/or failures. This book, looks at various angles of this problem from the perspective of potential attack models and recommends mitigating solutions to such attacks.

References

1. A. Abur and A. Gómez Expósito. *Power System State Estimation: Theory and Implementation*. CRC Press, Boca Raton, FL, USA, 2004.
2. P. L. Ambassa, A. Kayem, S. Wolthusen, and C. Meinel. Secure and reliable power consumption monitoring in untrustworthy micro-grids. In Robin Doss, Selwyn Piramuthu, and Wei ZHOU, editors, *Future Network Systems and Security*, volume 523 of *Communications in Computer and Information Science*, pages 166–180. Springer, Cham, Switzerland, 2015.
3. P. L. Ambassa, S. Wolthusen, A. Kayem, and C. Meinel. Robust snapshot algorithm for power consumption monitoring in computationally constrained micro-grids. In *Smart Grid Technologies - Asia (ISGT ASIA), 2015 IEEE Innovative, Bangkok, Thailand*, pages 1–6, Piscataway, NJ, USA, 3–6 Nov. 2015. IEEE Press.
4. A. Baiocco, S. Wolthusen, C. Foglietta, and S. Panzieri. A model for robust distributed hierarchical electric power grid state estimation. In *Innovative Smart Grid Technologies Conference (ISGT), 2014 IEEE PES*, pages 1–5, Piscataway, NJ, USA, Feb 2014. IEEE Press.
5. P. Buchana and T. S. Ustun. The role of microgrids amp; renewable energy in addressing sub-saharan Africa's current and future energy needs. In *Renewable Energy Congress (IREC), 2015 6th International*, pages 1–6, Sousse, Tunisia, 24–26 March 2015. IEEE Press.
6. Y. Feng, C. Foglietta, A. Baiocco, S. Panzieri, and S. Wolthusen. Malicious false data injection in hierarchical electric power grid state estimation systems. In *Proceedings of the Fourth International Conference on Future Energy Systems*, e-Energy '13, pages 183–192, New York, NY, USA, 2013. ACM.
7. K. Iniewski. *Smart Grid: Infrastructure and Networking*. McGraw Hill, New York, NY, USA, 2013.

8. A. Kayem, C. Meinel, and S. Wolthusen. A smart micro-grid architecture for resource constrained environments. In *2017 IEEE 31st International Conference on Advanced Information Networking and Applications (AINA)*, pages 857–864, March 2017.
9. G. N. Korres. A distributed multiarea state estimation. *IEEE Transactions on Power Systems*, 26(1):73–84, Feb 2011.
10. R. Kuwahata, N. Martensen, T. Ackermann, and S. Teske. The role of microgrids in accelerating energy access. In *3rd IEEE PES Innovative Smart Grid Technologies Europe (ISGT Europe)*, pages 1–9, Piscataway, NJ, USA, Oct 2012. IEEE Press.
11. Z. Liu. Chapter 3 - a global energy outlook. In Zhenya Liu, editor, *Global Energy Interconnection*, pages 91–100. Academic Press, Boston, 2015.
12. A. M. C. Marufu, A. Kayem, and S. Wolthusen. A distributed continuous double auction framework for resource constrained microgrids. In *10th International Conference on Critical Information Infrastructures Security (CRITIS 2015), October 5–7, 2015, Berlin, Germany*, pages 183–196. Vol. 9578, Lecture Notes in computer Science, Springer, 2015.
13. A. M. C. Marufu, A. Kayem, and S. Wolthusen. Fault-tolerant distributed continuous double auctioning on computationally constrained microgrids. In *2nd International Conference on Information systems Security and Privacy (ICISSP 2016), February 19–21, 2016, Rome, Italy*, pages 448–456. SCITEPRESS, 2016.
14. A. M. C. Marufu, A. V. D. M. Kayem, and S. D. Wolthusen. Power auctioning in resource constrained micro-grids: Cases of cheating. In Grigore Havarneanu, Roberto Setola, Hypatia Nassopoulos, and Stephen Wolthusen, editors, *Critical Information Infrastructures Security*, pages 137–149. Springer International Publishing, Cham, 2017.
15. E. D. Moe and A. P. Moe. Off-grid power for small communities with renewable energy sources in rural Guatemalan villages. In *Global Humanitarian Technology Conference (GHTC), 2011 IEEE*, pages 11–16, Piscataway, NJ, USA, Oct 2011. IEEE Press.
16. D. Nikolaev, Nikovski, Z. Wang, A. Esenther, H. Sun, K. Sugiura, T. Muso, and K. Tsuru. Smart meter data analysis for power theft detection. In Petra Perner, editor, *Proceedings of the 9th International Conference on Machine Learning and Data Mining in Pattern Recognition (MLDM 2013)*, volume 7988 of *Lecture Notes in Computer Science*, pages 379–389, New York, NY, USA, Jul 2013. Springer.
17. T. Winther. Electricity theft as a relational issue: A comparative look at Zanzibar, Tanzania, and the Sunderban Islands, India. *Energy for Sustainable Development*, 16(1):111–119, 2012.
18. G. K. Weldehawaryat, P. L. Ambassa, A. M. C. Marufu, S. D. Wolthusen, and A. Kayem, "Decentralised scheduling of power consumption in micro-grids: Optimisation and security," in *Security of Industrial Control Systems and Cyber-Physical Systems - Second International Workshop, CyberICPS 2016, Heraklion, Crete, Greece, September 26–30, 2016, Revised Selected Papers*, vol. 10166 of *Lecture Notes in Computer Science*, (Heraklion, Greece), pp. 69–86, Springer, Sept 2016.
19. P. Vytelingum, S. D. Ramchurn, T. D. Voice, A. Rogers, and N. R. Jennings, "Trading agents for the smart electricity grid," in *9th International Conference on Autonomous Agents and Multiagent Systems (AAMAS 2010), Toronto, Canada, May 10–14, 2010, Volume 1-3*, pp. 897–904, International Foundation for Autonomous Agents and Multiagent Systems, 2010.

Chapter 2
A Review on Attacks and Their Countermeasures in Power System State Estimation

Ammara Gul and Stephen D. Wolthusen

Abstract In this chapter we seek to provide an integrated, up-to-date survey of various attack models and corresponding protection regarding state estimation in larger-scale power systems. After giving brief overview of numerous attack and defence strategies in literature, the most appropriate between them are reported, explored and compared. Comparisons are performed on account of complexity level, optimality, considered structure, topology knowledge requirements and practical implementation along with the impact of advanced metering devices.

Keywords Power systems · State estimation · Attacks · Bad data · Mitigation

2.1 Introduction

Modern societies rely heavily on continuous operation of power systems, which in turn rely on information technologies for their efficient and safe operation. At present many power networks are being transformed into smart grids to maximise efficiency, which places new demands for timeliness and accuracy on power network state estimation compared to slower state estimation cycles used in conventional grids. As shown in Fig. 2.1, state estimation relies on measurements from a necessarily incomplete set of measurement points and topology information that itself may be incomplete and subject to topology analysis; crucial tasks such as

A. Gul (✉)
Department of Mathematics and Information Security, Royal Holloway University of London, Egham, Surrey, UK
e-mail: Ammara.Gul.2015@live.rhul.ac.uk

S. D. Wolthusen
Department of Mathematics and Information Security, Royal Holloway, University of London, Egham, Surrey, UK

Norwegian Information Security Laboratory, Gjovik University College, Norwegian University of Science and Technology, Trondheim, Norway
e-mail: stephen.wolthusen@rhul.ac.uk

A. V. D. M. Kayem et al. (eds.), *Smart Micro-Grid Systems Security and Privacy*,
Advances in Information Security 71, https://doi.org/10.1007/978-3-319-91427-5_2

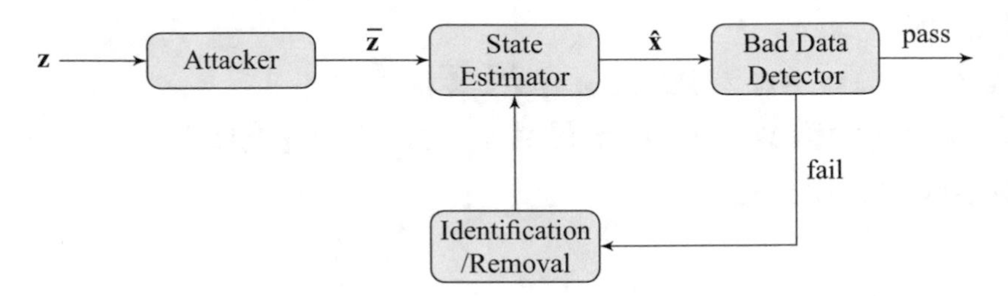

Fig. 2.1 State estimation model with attacker and detection test [3]

contingency analysis must rely on results of state estimation [1]. To operate such large, interconnected system reliably, we need information structure to be secure and effective against both failures and attacks. Failures can, e.g. be caused by a faulty measurement from a meter or operator's random error, but can also be caused by malicious activity. This survey primarily aims at considering the second cause of failures.

State estimation has attracted the attention of many research groups and there is a formidable volume of literature produced in last few decades. Attack strategies and their mitigation also gained much significance in the research community. Although each state estimator is armed with efficient detection schemes, these are only designed to prevent random errors from propagating into the power system. Coordinated attacks can be designed so as to avoid detection. This insight—albeit already formulated at the time state estimation was originally proposed for power networks—gave rise to the investigation of false data injection attacks by Liu et al. in [2] and classes of data injection attacks and their detection. Based on the requirements such as topology, infrastructure, and synchrophasor placement, we highlight a number of critical lines of research. We have chosen to limit the scope of work discussed somewhat to lines of research that are likely to be implementable in the short and medium term, neglecting proposals for mitigation that would require major changes to power network structure and instrumentation.

The remainder of the chapter is structured as follows. In Sect. 2.2, state estimation and bad data detection and removal procedures are reviewed. Section 2.3 presents the developments made under data attacks and their mitigations along with a descriptive comparison. Finally, concluding remarks and future work are discussed in Sect. 2.4.

2.2 Power System State Estimation

A state estimator evaluates the most likely state of the system by filtering and processing the measurements from Remote Terminal Unit (RTU) installed on transmission lines. There are two well known methods to solve the state estimation problem which are Weighted Least Square (WLS) Method and Weighted Least Absolute Value (WLAV) Method. Although WLAV is robust and stable in the sense that it is able to reject bad data efficiently it has some major drawbacks. It involves time consuming Linear Programming (LP) techniques, convergence rate reduces due to inclusion of auxiliary variables while minimizing and it is not reliable when encountered with leverage points (i.e., ill-conditionality may occur). Therefore, WLS (although not overly effective in the presence of bad data) is considered as the most widely used method to solve SE problems. For more details on WLAV, please see [4]. The WLS problem involves solving a non-linear set of equations relating measurements and state variables (voltage magnitudes and phase angles) by minimising the summation of squares of residuals.

2.2.1 Observability

Before going into the process of state estimation, it is necessary to define observability.

Definition 2.1 A continuous-time system

$$\dot{\mathbf{x}} = \mathbf{A}\mathbf{x} + \mathbf{B}\mathbf{u}$$

$$\mathbf{y} = \mathbf{C}\mathbf{x}$$

is *observable* if for any initial state $\mathbf{x}(0)$ and any final time $t > 0$ the initial state $\mathbf{x}(0)$ can be uniquely determined by knowledge of the input $\mathbf{u}(\tau)$ and output $\mathbf{y}(\tau)$ for all $\tau \in [0, t]$.

Definition 2.2 The n-state continuous linear time-invariant system defined above has the observability matrix $\mathbf{Q}$ defined by

$$\mathbf{Q} = \begin{bmatrix} C \\ CA \\ \vdots \\ CA^{n-1} \end{bmatrix}$$

The system is observable if and only if $\rho(\mathbf{Q}) = n$, where ρ is the row rank of a matrix.

This implies that a system is said to be observable if there exist a solution to Eq. (2.5) or we can say, a necessary and sufficient condition of observability is the existence of full rank Jacobian matrix $\mathbf{H}$ i.e., for each state variable, there exists at least one measurement. In case of non-observability, the observable islands are determined, and then, pseudo-measurements are included to resume observability.

2.2.2 State Estimation

Consider the non-linear AC power flow model as,

$$\mathbf{z} = h(\mathbf{x}) + \mathbf{e} \tag{2.1}$$

where $\mathbf{z}$ is the vector of measurements (m vector), $\mathbf{x}$ is the state vector (n vector, and $m > n$), $h(.)$ is usually a non-linear function relating measurements to the states and $\mathbf{e}$ is the vector of measurement errors having zero mean and known co-variance, which is denoted by $\mathbf{R}$. The errors are assumed to be independent, therefore, $\mathbf{R}$ is a diagonal matrix.

$$Cov(\mathbf{e}) = \mathbf{R} = diag\{\sigma_1^2, \sigma_2^2, \cdots, \sigma_m^2\}$$

Now, the objective function will be given as

$$J(\mathbf{x}) = \sum_{i=1}^{m} (\mathbf{z}_i - h_i(\mathbf{x}))^2 / \mathbf{R}_{ii} = [\mathbf{z} - h(\mathbf{x})]^T \mathbf{R}^{-1} [\mathbf{z} - h(\mathbf{x})]$$

which is to be minimized and the first order optimality condition is

$$g(\mathbf{x}) = \frac{\partial J(\mathbf{x})}{\partial \mathbf{x}} = -\mathbf{H}^T(\mathbf{x})\mathbf{R}^{-1}[\mathbf{z} - h(\mathbf{x})] = 0 \tag{2.2}$$

Where $\mathbf{H}(\mathbf{x}) = \partial h(\mathbf{x})/\partial \mathbf{x}$. After expanding $g(\mathbf{x})$ with Taylor series and writing the relation of $k+1$ iteration in terms of kth iteration

$$\mathbf{x}^{k+1} = \mathbf{x}^k - \mathbf{G}(\mathbf{x}^k)^{-1} g(\mathbf{x}^k) \tag{2.3}$$

Where $\mathbf{G}(\mathbf{x})$ is the Gain matrix

$$\mathbf{G}(\mathbf{x}^k) = \frac{\partial g(\mathbf{x}^k)}{\partial \mathbf{x}} = \mathbf{H}^T(\mathbf{x}^k)\mathbf{R}^{-1}\mathbf{H}(\mathbf{x}^k) \tag{2.4}$$

With the help of the above three equations, the normal equation to solve the state estimation problem will be

$$\mathbf{G}(\mathbf{x}^k)\Delta\mathbf{x}^{k+1} = \mathbf{H}^T(\mathbf{x}^k)\mathbf{R}^{-1}(\mathbf{z} - h(\mathbf{x}^k)) \tag{2.5}$$

Where $\Delta\mathbf{x}^{k+1} = \mathbf{x}^{k+1} - \mathbf{x}^k$.
We can summarize this method by a simple algorithm.

- Initialize the state vector $\mathbf{x}^k$ for $k = 0$ and get the measurement function $h(\mathbf{x})$.
- Calculate the Jacobian $\mathbf{H}(\mathbf{x})$ and the gain matrix $\mathbf{G}(\mathbf{x})$ from Eq. (2.4).
- Determine the right hand side of normal Eq. (2.5) and solve it for $\Delta\mathbf{x}^k$

- Check for convergence, $| \Delta \mathbf{x}^k | \leq \epsilon$.
- If not converged, update $\mathbf{x}^{k+1} = \mathbf{x}^k + \Delta \mathbf{x}^k$ and get a new $h(\mathbf{x})$ and repeat the above procedure.

It is assumed that now the reader has a sufficient understanding of state estimation (see reference [1] for details).

Even though model (2.1) is non-linear, generally the state is estimated by iteratively solving linearized WLS state estimation considering the steady state model. Therefore, to analyse the impact of bad data attack on state estimation, DC-model is adopted [1] rather than (2.1) and can be written as

$$\mathbf{z} = \mathbf{H}\mathbf{x} + \mathbf{e} \tag{2.6}$$

where $\mathbf{H}$ is the measurement matrix and each row of $\mathbf{H}$ corresponds to the type of measurement. For instance, for noiseless measurements $\mathbf{z} = \mathbf{H}\mathbf{x}$ and $\mathbf{z}_k = B_{ij}(\mathbf{x}_i - \mathbf{x}_j)$ denotes the line flow from bus i to j, where B_{ij} is the line susceptance and $\mathbf{x}_i, \mathbf{x}_j$ are voltage phase angles for buses i and j respectively [1]. Hence the set of normal equations is reduced to

$$\mathbf{x}^{k+1} = (\mathbf{H}^T(\mathbf{x}^k)R^{-1}\mathbf{H}(\mathbf{x}^k))^{-1}\mathbf{H}^T(\mathbf{x}^k)\mathbf{R}^{-1}\mathbf{z}^k \tag{2.7}$$

2.2.3 *Bad Data Analysis*

An important task after estimating the state is bad data analysis and hence, this function involves the test on the normalized residuals defined as

$$\mathbf{r}_{Ni} = \frac{\mathbf{r}_i}{\sqrt{var(\mathbf{r}_i)}} = \frac{\mathbf{z}_i - h_i(\mathbf{x})}{\sqrt{\mathbf{R}_{ii}}} \tag{2.8}$$

for the i-th measurement where $i = 1, \cdots, m$. A threshold is set for the normalized residual, and the above test is performed for each iteration. Normalized residual values larger than the fixed threshold are detected, and corresponding measurements are flagged as bad, and after their removal, state estimation can be re-run until all the bad data are removed, and the system converges. This basic and widely used detection test in WLS is known as largest normalised residual test. There are other testing schemes as well such as, χ^2-test or hypothesis testing identification (HTI) (Please see [1] by Abur for more details). Numerical stability refers to the impact of an incorrect/false input on the execution algorithm, therefore for a sound state estimation, our estimators must be numerically stable (although it is not the case always especially while using WLS state estimator [5]).

Bad data analysis is the ability of the state estimator to reject the erroneous measurements. For instance, if there exist some faulty meter or the bad data induced by an attacker, Eq. (2.1) can be written as

$$\mathbf{z} = h(\mathbf{x}) + \mathbf{e} + \mathbf{a} \tag{2.9}$$

where $\mathbf{a}$ denotes, the induced bias or bad data and its detection/identification is known as bad data analysis. Bad data can be injected by different ways by an adversary i.e., either by modifying some meter measurements by physically tampering the meters or by compromising some transmission line(s).

The process of state estimation is still the same as the one given in the above algorithm. Before going into the details of different attack and mitigation schemes, some definitions are worth mentioning:

Definition 2.3 Supervisory Control And Data Acquisition (SCADA) is a system for remote monitoring and control that operates with coded signals over communication channels (using typically one communication channel per remote station).

Definition 2.4 Remote Terminal Units (RTUs) is a microprocessor-controlled electronic device that interfaces objects in the physical world to a distributed control system or SCADA (supervisory control and data acquisition) system by transmitting telemetry data to a master system, and by using messages from the master supervisory system to control connected objects.

Definition 2.5 Intelligent Electronic Devices (IEDs) are the devices incorporating one or more processors with the capability to receive or send data/control signals from or to an external source (e.g., electronic multi function meters, digital relays, controllers) [6].

Definition 2.6 Phasor Measurement Units (PMUs) are the devices that measure voltage and current magnitudes using a global positioning system (GPS) reference source for synchronization with an accuracy of $1\,\mu$s. The resultant time tagged phasors can be transmitted to a local or remote receiver at rates up to 60 samples per second [6].

2.3 Descriptive Comparison Between Various Strategies

For the notational harmony, the symbols in some papers have been moderately adjusted. As a reference point, we use the formulation of [7] which we followed in Sect. 2.2 as well. There are some common features upon which this section relies, for instance, the type of model considered is steady state linearised DC state estimation model and the class of attacks studied is data driven attacks unless otherwise specified. We define two attack categories upon which the following papers discuss; injection and non-injection attacks. Injection attacks include false data injection in meters (Sect. 2.3.1), data injection on SCADA systems (Sect. 2.3.3), false data injection in limited number of meters (Sect. 2.3.4), injection in just one control centre (Sect. 2.3.5), attacks with multiple adversaries (Sect. 2.3.10) and attacks that are successful despite detection (Sect. 2.3.8). Non-injection attacks include swapping or replaying attacks (Sect. 2.3.6), delay attacks (Sect. 2.3.7), jamming or suppression of measurements (Sect. 2.3.9). Soon after the introduction of false data injection attacks by Liu et al., Bobba et al. presented a notable protection scheme stated in Sect. 2.3.2.

2.3.1 Origin of False Data Injection Attacks

Substations are generally equipped with meters whether the traditional RTUs or the advanced IEDs. Meters in a power system can be attacked or the computer systems in which all the measurements are stored can be hacked. As a consequence, the resulting bad measurements can cause cascading failures if not detected and identified in polynomial time. With the realization of the hazardous impacts of bad data, power systems researchers developed various methods to cope with it [7]. These methods detect, identify and remove the erroneous bad data. Generally, these techniques use *Largest Normalized Residual Test (LNRT)* which states that, "when bad measurements take place, the squares of difference between the observed measurements and their corresponding estimates often become significant" [2]. In a novel approach proposed by Liu et al., to orchestrate a type of coordinated attack that can thwart the conventional detection schemes by dodging the system operators. Such attacks are called *false data injection (FDI) attacks* or stealthy/hidden attacks [2].

2.3.1.1 Description

A power network is considered with m meters providing m measurements. There are n state variables associated with them by a measurement function h as given in Eq. (2.1). Firstly, it is proved that if the attack vector can be designed by combining linearly the columns of the Jacobian matrix $\mathbf{H}$ (already defined in section II) or $\mathbf{a} = \mathbf{Hc}$, it can evade the detection scheme (provided earlier). On this basis, Liu et al. examined two *attack goals* [2]:

(1) *Random False Data Injection Attacks*: The attacker intends to find any attack vector satisfying the above condition. (2) *Targeted False Data Injection Attacks*: The attacker attempts to determine a specific attack vector to force certain state variables to be fallacious. It can be deduced that while the former are easier to perform, the latter are more damaging [2]. Under targeted FDI attacks, further two cases are observed: unconstrained case: in which attacks are designed to manipulate certain state variable irrespective of their impact on others and constrained case: in which attacks are performed to compromise certain variable while keeping others unaffected.

Two logical attack scenarios are taken into account: In *Scenario I: Limited access to meters*, the attacker is constrained to attack only a particular set of measurements mainly due to the higher physical security. In *Scenario II: Limited resources available to compromise measurements*, the attacker is restricted to some specific number of readings to attack due to the finite resources. These attacks are also extended to the notion of generalized false data injection attacks. Simulations proved their capability of launching stronger attacks than the false data injection attacks. However, these attacks are restricted by real world constraints and therefore not very threatening at the moment.

2.3.1.2 Discussion

- Attacker is assumed to have the knowledge the system topology or the Jacobian matrix **H** [2]. It is an open question for future research to orchestrate false data injection attacks with partial topology knowledge or even with no knowledge of **H** [8–10].
- For theoretical observation, a linear model state estimation is presented in [2] while simulations to verify the possibility of the attacks are shown on non-linear models based on SCADA/EMS (Energy Management System) test bed. Therefore, modelling non-linear state estimator for false data injection attacks can be one of the future work directions.
- Also, the simulations tested each scenario for only 10^3 times on a random basis leaving some space unexplored [2]. The optimal solution to which measurements to be manipulated is almost unknown. Following this, there are multiple papers [11, 12] on this notion.

2.3.2 *Immunity by Protecting Critical Measurements*

While Liu et al. presented the stealthy attack strategies from the attacker's point of view and demonstrated what the attacker require to perform an attack without being detected, Bobba et al. looked at the problem from the system operator's perspective. Bobba et al. in [13], demonstrated a practical scheme instead of providing new algorithms for detection. It is proved that after *protecting a particular set of measurements* (equal to the number of state variables), the system can be made immune to false data injection. On this basis, it is quite useful to protect the whole system only by protecting a small set of transmitted variables. An alternative can be independent verification of certain carefully chosen state variables or it is also possible for the operator to benefit.

2.3.2.1 Description

The mathematical model to determine the measurements to be protected is as follows [13]: Let M be the set of measurement indices and $I_{\bar{m}}$ denote the indices of protected measurements while I_m denoting its complement. Similarly, let V be the set of indices of state variables and $I_{\bar{v}}$ denote the indices of independently verified state variables while its complement is I_v. For any stealthy attack vector $\mathbf{a}_i = \mathbf{Hc}_i$, if $i \in I_{\bar{m}}$ or $i \in I_{\bar{v}}$, $\mathbf{a}_i = 0$ which implies that attacker cannot find such stealthy attack vector $\mathbf{a}$. It is now clear that the system operators have to protect the minimum $I_{\bar{m}}$ or $I_{\bar{v}}$ in order to make the system secure. The problem now is to find identify the optimal $I_{\bar{m}}$ and $I_{\bar{v}}$. Two approaches are used:

Approach I Brute-Force Search This attempt to determine $I_{\bar{m}}$ and $I_{\bar{v}}$ is a straight forward brute-force approach. Let $p =\mid I_{\bar{m}} \mid$ or $q =\mid I_{\bar{v}} \mid$ where $0 \leq p \leq m$ and $0 \leq q \leq n$. System operators need to search from $\binom{m}{p} * \binom{n}{q}$ combinations for their choice of p and q to find the optimal sets such that after protecting, no stealthy attack is possible.

Approach II Protecting Basic Measurements A set of basic measurements in state estimation is a minimum set of measurements which makes the system observable [13]. It is evident that cardinality of such set must be n. In this approach, basic measurements are protected by defining an equivalent mapping by which all measurements are identified either as basic or as redundant. Some adaptations of these methods with some modifications can be seen in [12].

2.3.2.2 Discussion

- Note that, the set of basic measurements is equal in number to the state variables i.e., n. Protecting such a large set is not a feasible approach to make the system resilient in case of attacks due to cost and time constraints [12]. Being an active research area, numerous researchers approach this problem with the construction of greedy algorithms [11, 14].

2.3.3 Minimum Cost Stealth Attacks

In [12], it is proved that *stealthy attacks can be launched on SCADA systems* and two security indices α_k and β_k are formulated defining sparse attacks and small magnitude attacks respectively. Hence, in [15] the security index α_k quantifying the minimum measurements to be modified in performing the successful attacks (or minimum cost attacks) is computed.

2.3.3.1 Description

The research considered by Dan and Sandberg [15] is threefold, firstly, a security index α_k for minimum cost attack is computed. It is the least number of measurements that need to be manipulated to perform a specific attack. Then, a partitioning of the set of measurements is defined in such a way that a cluster of measurements is available to attacker at the unit cost. This can be the case when attack is performed from a substation and technically, all of its measurements can be attacked at once. Finally, protection schemes are devised utilizing the same cluster strategy. Two approaches are defined for this purpose:

1. *Perfect protection:* is the set of protected measurements P such that no stealthy attack is possible. It has further two categories, i.e., protecting stealthy meter attacks and protecting RTUs from attacks. The cost of perfectly securing all measurements from attacks is quite expensive, and cost is equal to $n = \mid P \mid$. Whereas, we can find a dominating set of RTUs such that no RTU is vulnerable to stealthy attack with the cost much less than n.
2. *Non-perfect protection:* is the set of maximal secured measurements such that stealthy attacks can be minimum. For this purpose, two possible metrics are inspected i.e., maximal minimum attack cost and maximal average attack cost [15]. For the former, the operator intends to maximize the minimum attack cost for all the measurements that can possibly be attacked and for the later one, the operator aims to maximize the average minimum cost for the likely attackable measurements. For both, simple greedy algorithms that aim to find the optimal solution can be leveraged.

2.3.3.2 Discussion

- Perfect protection against stealthy meter attacks is out of the question as it requires atleast n measurements to be protected (same as provided by Bobba) while for the security against RTU attacks, *Dominating Set Augmentation Algorithm(DSA)* is used with initiating the set of protected measurements P with a minimal dominating set rather than a flat start. One research question might be giving a flat start to see the efficiency of the algorithm.
- For non-perfection protection, high-level redundancy is required by both greedy algorithms to reach the optimal solution. Results are quite favourable but no argument on the convergence time is made. Resilience and time limitations of greedy algorithms by lowering the redundancy might be tested in future.

2.3.4 Sparse Attacks Corrupting Two Injection Meters

A discrepancy in [2] is rectified as the scheme did not answer which measurements to be compromised. In [16] by Giani et al., a resilient algorithm is proposed to determine (≤ 5) sparse attacks involving only two power meters and arbitrary line meters. Precisely 3, 4 *and* 5 *sparse attacks* can be devised when all lines are metered.

2.3.4.1 Description

Failures occur in the power system mainly because of either some faulty component/meter or due to some malicious activity that leaves the system unobservable, and these unobservable stealthy attacks need coordination to evade detection. Here, in [16] low-sparsity stealthy attack is under consideration that requires coordination of at most 5 m. Stealthy attacks involving large number of meters are uncertain because of the level of coordination required to perform them. An effective algorithm is proposed by Giani et al. in [16] to determine all the possible unobservable attacks that require only two injection meters and arbitrary line meters to be manipulated with $O(n^2m)$ computational complexity with n buses and m line meters. In the specific situations when there are meters on each line, canonical forms for 3, 4 and 5 sparse unobservable attacks can be derived by the algorithms from graph theory requiring complexity of $O(n^2)$ to determine the possibility of these canonical forms.

As far as the detection of stealthy unobservable attacks (not compulsory to be sparse) are concerned, utilizing the known-secure PMUs is proposed [16]. Location of PMUs is identified by buses at which PMUs must be installed to thwart the stealthy attacks. Problem of determining the minimal sufficient PMUs is NP-hard, therefore it is verified that by placing $p + 1$ known secure PMUs, system can be protected from P unobservable attacks. An efficient algorithm to find this placement has complexity of $O(n^2 p)$.

2.3.4.2 Discussion

- Although PMUs are considered to be the most reliable source of data base as they use GPS to provide synchronous measurements. Here, known secure PMUs are assumed to ignore any fault occurred in PMUs or in GPS that provide the time-stamped signals. For example, GPS spoofing attacks in [8] can cause extensive damage to the power systems hence, unfolding numerous future research questions.
- Another yet interesting question is to explore the applicability of these models on decentralised state estimation.

2.3.5 *Stealth Attacks Involving Exactly One Control Centre*

False data injection attacks are also possible in decentralised state estimation structure [17]. Ognjen and Gyorgy in [18] presented five attack strategies for *distributed state estimation* (DSE) provided the attacker require the knowledge of the system topology. Further, an attack involving a single control centre is considered in DSE that seems to be successful in either divergence or the erroneous convergence of the system. A similar approach [19] in which attacker aims to compromise the infrastructure of a single control centre.

2.3.5.1 Description

The work [19] is an equivalent version of the previous work of Vukovic and Dan in [18]. Additionally, stealthy attacks on the fully distributed state estimation is being considered for the first time. In the former, stealthy attack that requires the corruption of a single control centre is examined whereas in this paper, authors focussed on the manipulation of the communication infrastructure of a single control centre. Byzantine consensus problem is considered as a baseline in which there are processors that have to consent on a single value even if an error is reported by a processor. In this work, regions act as processors but attack is different. Therefore, resulting in a successful *denial of service* attack i.e., this attack can blind the system operators of the individual area. First singular vector (FSV) attacks and uniform rotation (UR) attacks are applied, and it is verified that even small FSV attacks can cause the desired damage when the state estimation converges with a minimum of 10% error.

Also, an efficient and novel mitigation scheme that not only support convergence but also let the attack to be localized. Starting with the token assumption that every region uses to express their beliefs. Empirical frequency (of token visits) is evaluated for every region. A high empirical frequency determine the likely corrupted region. Exploiting Markov chain to model this random walk of a global observer, the belief consensus localization algorithm (BCL) for regional operators is proposed. Any compromised region is identified and after isolating the infected region, state estimation is re-run until convergence.

2.3.5.2 Discussion

- Although the attacker is not assumed to have access to all entries of $\mathbf{H}$ rather it knows the estimate of the previous iteration which is helpful in launching the attack. The subject of the future work is to study the impacts while alleviating this requisite.
- Numerical results proved the argument of both the attack performance and their diminution. However, it can be seen that smaller weak attacks can not be detected in polynomial time that can make the convergence fallacious as can be observed in the first part of [19]. Hence, a fair research might examine this in future.

2.3.6 *Re-ordering or Swapping Attacks*

Unlike the work mentioned before, most attacks, however, rely on the assumption that arbitrary values may be injected by an adversary. It is argued that this may not be a realistic assumption and that instead, it is of considerable interest to study cases where measurements and communication channels are protected, at least using authentication and integrity protection as provided, e.g. by the ISO/IEC 62351 standard. This offers a more realistic adversary model compared to that introduced by Liu et al. [2].

2.3.6.1 Description

A novel attack relying solely on re-ordering or replaying of the measurement vector which result in spurious estimates is proposed in [20]. Here, we formulated targeted re-ordering attack considering two scenarios for this: (1) swapping the measurements by the previous plausible vector and (2) swapping the measurements by some scalar multiple of previous measurement vector. For both cases, two security metrics are introduced one for the sparse re-ordering attack and the other for small magnitude re-ordering attack. We proved that for scenario I, if the attacker swaps more than 80% of total measurements, it can cause the system to diverge as a result of ill-conditioned Jacobian. Similarly, to execute attacks of the kind as in scenario II, attackers have to pay more (swapping of about 75% measurements will be required) to get maximum mean square error in estimated states.

2.3.6.2 Discussion

- Currently in our power grid, the measurements are not authenticated time-stamped to detect such re-ordering and such authentication for detection purposes is adequately expensive to implement atleast till near future.
- Even assuming time-stamped authentication, which is offered by ISO/IEC 62351 but not widely deployed at present, re-ordering attacks may still succeed when combined with message spoofing. This implies that as long as there are old components in our power network, there can be a chance of these kind of attacks.

2.3.7 *Random and Structured Delay Attacks*

It is assumed that installing PMUs is the most genuine solution to stealthy attacks. But Shepard and Humphreys, in [8], introduced *GPS spoofing attacks* that has the ability to change the measurement of PMUs just by delaying the signal for some μs. In the last decade, civil GPS spoofing is becoming a serious threat to smart grids which are heavily relied on PMUs. On the argument of GPS spoofing attacks, Baiocco et al. defined *random and structured delay* attacks in HSE [9]. In these kind of attacks, adversary do not require the complete knowledge of the topology, and with very few trivial assumptions, severe impacts in form of ill-conditionality of the Jacobian matrix or instability of power systems can be observed.

2.3.7.1 Description

On hierarchical state estimation, Baiocco et.al in [9] exploited delay and jitter attacks considered with their possible applicability on CSE and DSE as well with very low constraints. Three-level hierarchy is proposed where, either top-down or bottom-up synchro-upgrade procedure is followed. In either case, estimated states of every level has to pass on to the next in a synchronous manner so that the whole system state can be evaluated in time upon which contingency analysis heavily rely.

With the introduction of delay between the levels, stealthy attacks are possible. Two types of attacks are examined, (1) Random delay (jitter) and (2) Structured delay (jitter). It can be observed that while random delay/jitter attacks are easier to perform, the structured have more adverse impacts. These attacks produce strong outcomes in the form of ill-conditionality of the Jacobian matrix or instability of the power system. Majority of the above mentioned stealthy attacks need to be coordinated to avoid detection. For this purpose, system topology or Jacobian $\mathbf{H}$ in addition to the manipulated measurements or state variables is assumed to be known to the attacker. Surprisingly, to launch delay or jitter attacks, its not necessary for the attacker to have the in-depth knowledge of the topology.

2.3.7.2 Discussion

- Random delays require no prior knowledge of the system (in depth) and therefore are less effective than structured delays with the assumption of known Jacobian $\mathbf{H}$. One might examine impacts of small structured delay attacks with partial or no knowledge of $\mathbf{H}$ for future study.
- An active research area for further research might be on the mitigation policies for the delay and jitter attacks (for both random and structured).
- In the future smart grid power systems, these attacks might be explored in fully distributed state estimation.

2.3.8 Subspace Methods for Data Attacks

Major part of previous work on the security of power system state estimation focus on stealthy attacks that avoid bad data detection tests. To our knowledge, *data framing attacks* by Kim et al. is the first piece of work towards detectable attacks that proves to be successful despite detection by misleading the error identifier [3]. *Subspace methods* for constructing data framing attacks have recently been formulated in [21] while assuming that the attacker is only capable of manipulating a subset of the measurement vector without the detailed knowledge of **H** or the system parameters.

2.3.8.1 Description

The research by Kim et al. in [21] is twofold: firstly, unobservable data attacks are designed with the help of subspace methods with only partial measurements and secondly, subspace information, is used to orchestrate data framing attacks with the similar requirement of partial measurements.

All information that an attacker require is the subspace of **H** i.e., $R(H)$. Two algorithms are proposed to perform successful data-driven attacks: (1) Attack with full measurements and (2) Attack with partial measurements. Due to similarity, we will discuss the latter while interested readers are advised to see [21] for details. Algorithm for data attacks with partial measurements is as follows:

Step 1 *Subspace Estimation:* Based on the available measurements, estimate the basis matrix U of $R(\mathbf{H})$ (subspace of **H**).

Step 2 *Null Space Estimation:* Calculate the null space of the matrix obtained by removing from basis matrix the rows corresponding to the critical set C just to ensure the non-attack positions.

Step 3 *Attack:* Corrupt the data from C by corresponding values of $\alpha.U$ where α being a scalar.

The subspace related data framing attacks exploit the bad data detection and removal techniques. Particularly, the attacker maximizes the residual of the framed measurements to trigger the false alarm purposely hence misguiding the system operator. After removing such data, despite of the consistency with the model, existing false measurements result in spurious estimates. Algorithm for data framing attack with partial measurements executes the same way as for unobservable data attacks given above (for details see [21]).

2.3.8.2 Discussion

- Majority of literature in countermeasures focus on protecting certain number of measurements to made the system un-attackable while assuming that the adversary has the knowledge of **H** or the system parameters. This paper opens many questions to rescale the mitigation and protection measures.
- On the other hand, it is revealed that today's power systems are not secure under these orthodox bad data detection and identification techniques. More work on bad data monitoring mechanism is required.

2.3.9 *Detectable Jamming Attacks*

Deka et al. in [22], later discovered that cardinality of the *detectable framing attacks* (introduced in previous subsection) can be reduced to more than 50% of the stealthy attacks by controlling the presence of certain protected measurements. Furthermore, the authors maximize the attack impact by the inclusion of *measurement jamming* into the detectable attacks [10].

2.3.9.1 Description

False data injection attacks are considered as unobservable when they remain undetected while testing through traditional schemes. All of the above mentioned work in Sect. 2.4 verifies adequate success of these stealthy/hidden attacks (with some assumptions) and their corresponding counter measures. But the concept of data framing detectable attacks in [3] have pressed the power security researchers. On this basis, [22] is the first known work (to our best knowledge) to examine the detectable false data injection attacks by Deka et al.. Following this, the authors present detectable jamming attacks by adding measurement jamming into it [10] to maximize the impact.

Earlier, it is proved by the same authors that the cardinality of detectable attacks can be reduced to more than half of that of stealthy attacks (or atleast half) [22]. In addition to performing detectable attacks, the adversary here is capable of jamming/blocking some measurements/communication in the network. Compared to bad data injection, jamming is less cost-intensive and therefore its cost varies from 0 to the maximum of P_d (where P_d be the cost of detectable attack without jamming). This way, jamming cost is partitioned into two regions to obtain the optimal attack by graph-theoretic means. One of the essential findings in this work is the ability of attacker to apply jamming only if the jamming cost is less than half of that of injection cost [10]. Since, determining the optimal detectable jamming attack is NP-hard, a polynomial time approximation is obtained to verify the results.

2.3.9.2 Discussion

- Ref. [10] is one of the most recent works (to our best knowledge) in this mention and protection against such attacks might be devised in near future to overcome the potential threats by detectable jamming attacks.
- In the perspective of an adversary, designing optimal detectable jamming attack must be the next task. In addition, data jamming in decentralised state estimation can be one of the areas of further study.

2.3.10 *Data Injection Attacks with Multiple Adversaries*

Till the end of 2015, almost all of the ongoing research focussed on investigating a kind of false data injection attacks involving a single attacker and examining the attack impacts on the security of the grid. Along with this, countermeasures are also proposed to cope with the mentioned class of attacks. Interestingly, no work on the notion of *multiple adversaries* is seen until Sanjab and Saad studied the impact of two attackers simultaneously [23].

2.3.10.1 Description

Following the above mentioned proposition for multiple attackers, the authors in [24] constructed two models from game theory relied on linearized/DC state estimation while considering centralized case. Successful attacks can manipulate the price and hence have financial benefits causing loss for the grid operators.

In the *first* model, Stackelberg game paradigm is used in which defender or the system operator act as a leader and the attackers as its followers. Thus, a non-cooperative game is played between the defender and the attackers noting that in this game, leader can predict the adversary's actions prior to playing its defence strategy (e.g., selecting the measurements to protect). Solution to this game is studied where defender needs to minimize the attack impacts and in parallel attacker chooses its strategy to maximize the trade off between benefits and attack cost. The only difference in the *second* game which is Nash equilibrium model (see Ref. [24] for details) is that now the defender can not anticipate the actions of adversaries and hence play to meet its certain objective regarding defence. In both of the mentioned paradigms, two situations can be observed: (1) The attackers can cancel the effect of each other resulting in no manipulation and hence no need to defend and (2) The attackers can help each other achieving their targets and therefore can be destructive for the grid.

2.3.10.2 Discussion

- Recently in Dec. 2015, Ukraine's power plant has been hacked so badly that the control centre operators had to manually operate the breakers for so many days following the attack. It is reported that the hack involved multiple adversaries [25].
- After this attack, it is also shown that the grids in the US are more vulnerable to these attacks as they have more automated breakers than Ukraine had. All of this call for more strategic defence of our power grids.

2.4 Conclusion

Last decade seems to be quite devoted in the study of false data injection attacks and their mitigation on power system state estimation. In this paper, we examined the most convincing of them. A general observation can be: false data injection attacks are a genuine threat to the power grids.

Two essential reviews after analysing above significant papers are: (1) Almost all of the above mentioned attacks proved their success against the weaker bad data detection test that relies on residuals (i.e., residual method). (2) Most of the papers considered the traditional WLS method for state estimation rather than using some better and advanced methods. Therefore, interactions with models other than WLS and residual method is an uncovered and open research question.

In Table 2.1 which overviews the comparison, one of the two quite conventional observations is the incompatibility of finding the optimal attack in which the attacker needs minimum knowledge of the system. Such ideal attacks are adequately impossible due to the real world constraints. Although, many researchers have succeeded to determine sub-optimal solutions that comes with acceptable but relatively low accuracy. Computational time required for such attacks along with the assessment of the encountered algorithmic complexity can be one of the further studies. Second point to be mentioned here is the lack of research in decentralized format and needs attention as with the up-gradation to power grid, decentralized structure is more likely.

Table 2.1 Table of comparison

Attack mechanism	Structure	Mitigation scheme	Optimality
False data injection attacks [2]	Centralized	...	Sub-optimal
Minimum cost stealth attacks [15]	Centralized	Algorithm to place encrypted devices	Sub-optimal
Sparse attacks with two injection meters [16]	Centralized	Known-secure PMUs	Sub-optimal
Stealth attacks involving exactly one control centre [19]	Decentralized	Markov chain based BCL-algorithm	Sub-optimal
Delay and jitter attacks [9]	Decentralized	...	Sub-optimal
Subspace based data attacks [21]	Centralized	...	Sub-optimal
Detectable jamming attacks [10]	Centralized	...	Sub-optimal
Attacks with multiple adversaries [24]	Centralized	...	Sub-optimal
Re-ordering or swapping attacks [20]	Centralized	...	Sub-optimal

References

1. A. Abur, A. G. Exposito: In: Power System State Estimation: Theory and Implementation. CRC Press - Taylor Francis Group ISBN: 9780824755706 Textbook - 327 Pages (March 2004)
2. Y. Liu, P. Ning, M. K. Reiter: False data injection attacks against state estimation in electric power grids. In: Proceedings of 16th ACM conference on Computer and communications security, NY, USA (November 2009) 21–32
3. J. Kim, L. Tong, R. J. Thomas: Data Framing Attack on State Estimation. IEEE Journal on Selected Areas in Communications **32** (2014)
4. M. Ahmed, Technology and Engineering. In: Power System State Estimation. Artech House (January 2013)
5. A. Baiocco, S. Wolthusen: Stability of Power Network State Estimation under Attack. In: Proceedings of 2014 IEEE Innovative Smart Grid Technologies - Asia (ISGT ASIA), Kuala Lumpur (May 2014) 441–446
6. A. Gomez-Exposito, A. Abur, A. de la Villa Jaen, C. Gomez-Quiles: A Multi Level State Estimation Paradigm for Smart Grids. Proceedings of the IEEE **99** (April 2011) 952–976
7. A. Monticellii: Business and Economics. In: State Estimation in Electric Power System: A generalized approach. Springer Science and Business Media (May 1999)
8. D. P. Shepard, T. E. Humphreys: Evaluation of the Vulnerability of Phasor Measurement Units to GPS Spoofing Attacks. In: Sixth annual IFIP Conference on Critical Infrastructure Protection. Volume 5., Washington DC (December 2012)
9. A. Baiocco, C. Foglietta, S. D. Wolthusen: Delay and Jitter Attacks on Hierarchical State Estimation. In: Proceedings of 2015 IEEE International Conference on Smart Grid Communications (SmartGridComm), Miami, FL, IEEE (November 2015) 485–490
10. R. B. D. Deka, S. Vishwanath: Optimal Data Attack on Power Grid: Leveraging Detection and Measurement Jamming. In: Proceedings of the 2015 IEEE International Conference on Smart Grid Communications (SmartGridComm), Miami, FL, IEEE (November 2015) 392–397
11. T. T. Kim, H. V. Poor: Strategic Protection Against Data Injection Attacks on Power Grids. IEEE Transactions on Smart Grid **2** (June 2011) 326–333
12. A. T. H. Sandberg, K. H. Johanasson: On Security Indices for State Estimators in Power Networks. In: Preprints of the First Workshop on Secure Control Systems CPSWEEK 2010, Stockholm (2010)
13. R. B. Bobba, K. Rogers, Q. Wang, H. Khurana, K. Nahrstedt, T. J. Overbye: Detecting false data injection attacks on DC state estimation. In: Proceedings of First Workshop on Secure Control Systems (SCS 2010), Stockholm, Sweden (April 2010)
14. S. Bi, Y. J. Zhang: Defending mechanisms against false data injection attacks in the power system state estimation. (December 2011)
15. G. Dan, H. Sandberg: Stealth Attacks and Protection Schemes for State Estimators in Power Systems. In: Proceedings of the 2010 first IEEE Smart Grid Communication, Gaithersburg, MD (October 2010) 214–219
16. A. Giani, E. Bitar, M. Garcia, M. McQueen, P. Khargonekar., K. Poolla: Smart grid data integrity attacks: characterizations and countermeasures. (October 2011)
17. Y. Feng, C. Foglietta, A. Baiocco, S. Panzieri, S. D. Wolthusen: Malicious False Data Injection in Heirarchical Electric Power Grid State Estimation Systems. (May 2013)
18. O. Vukovic, G. Dan: On the Security of Distributed Power System State Estimation under Targeted Attacks. (March 2013)
19. O. Vukovic, G. Dan: Security of Fully Distributed Power System State Estimation: Detection and Mitigation of Data Integrity Attacks. IEEE Journal on Selected Area Communication (July 2014)
20. A. Gul, S. D. Wolthusen: Measurement Re-Ordering Attacks on Power System State Estimation. (2017, Accepted in 7th IEEE International Conference on Innovative Smart Grid Technologies (ISGT), Europe)

21. J. Kim, L. Tong, R. Thomas: Subspace Methods for Data Attack on State Estimation: A Data Driven Approach. (March 2015)
22. D. Deka, R. Baldick., S. Vishwanath: Data Attacks on Power Grids: Leveraging Detection, IEEE (February 2015)
23. A. Sanjab, W. Saad: Smart Grid Data Injection Attacks: To Defend or Not? In: Proceedings of the 2015 IEEE International Conference on Smart Grid Communications (SmartGridComm), Miami, FL, IEEE (November 2015) 380–385
24. A. Sanjab, W. Saad: Data Injection Attacks on Smart Grids with Multiple Adversaries: A Game-Theoretic Perspective. Number 99 (April 2016)
25. K. Zetter: Inside the Cunning, Unprecedented Hack of Ukrain's Power Grid. Critical Infrastructures (March 2016)

Chapter 3
An Anonymous Authentication Protocol for the Smart Grid

Hikaru Kishimoto, Naoto Yanai, and Shingo Okamura

Abstract The Smart Grid allows users to access information related to their electricity usage via IP networks. Both the validity and the privacy of such information should be guaranteed. Consumers' electricity bills can then be charged directly to them via the Smart Grid, even outside their homes. Such information from this bill is strictly related to the privacy of consumers; hence, we propose an anonymous authentication protocol for electricity usage on the Smart Grid. Our main idea is to utilize group signatures with controllable linkability. In these group signatures, only designated signers can generate digital signatures with anonymity under a single group public key, and only entities with a link key can distinguish whether the signatures are generated by the same signer or not. Whereas our proposed protocol can include any group signature scheme with controllable linkability, we also propose new controllably linkable group signatures with tokens, which are handled by smart meters on the Smart Grid. We implement the proposed group signatures, and then estimate the computational time of our anonymous authentication protocol at about one-and-a-half seconds on Raspberry Pi.

Keywords Anonymous authentication · Authentication protocol · Group signatures · Privacy · Applied cryptography

---

H. Kishimoto (✉) · N. Yanai
Osaka University, Suita, Osaka, Japan
e-mail: hikaru.kishimoto@ist.osaka-u.ac.jp; yanai@ist.osaka-u.ac.jp

S. Okamura
National Institute of Technology, Nara College, Yamatokoriyama, Nara, Japan
e-mail: okamura@info.nara-k.ac.jp

A. V. D. M. Kayem et al. (eds.), *Smart Micro-Grid Systems Security and Privacy*,
Advances in Information Security 71, https://doi.org/10.1007/978-3-319-91427-5_3

3.1 Introduction

3.1.1 Background

The Smart Grid is a state-of-the-art digital information technology (IT) for an electricity infrastructure. According to the National Institute of Standards and Technology (NIST) [1], the Smart Grid allows for two-way communication between utility power grids and their customers with an IT infrastructure that uses advanced metering infrastructures (AMI). The Smart Grid consists of computers and controls, new technologies and equipment that connect to other networks and work together to give advantages to both the electrical utilities and their consumers. A smart meter at each home can check the status of electricity usage, manage its total amount in real time, and allow the meter to be read remotely. Although this undoubtedly gives advantage to consumers, one of the main concerns regarding the Smart Grid is security threats [2]. Since its system is connected with other networks, there are potential risks such as the manipulation of control information. Likewise, privacy concerns exist [3, 4] where sensitive consumer information may be revealed and compromised. The government of Netherlands, for example, rejected the mandatory usage of smart meters due to privacy concerns [5].

However, in spite of such concerns, electricity usage services exist whereby a consumer is allowed to pay an electricity bill that utilises outlets in public places. For example, when the battery of an electric vehicle is empty and its owner, i.e. a consumer, needs to power up the vehicle again, a manager of outlets that the consumer wants to utilise will not want the consumer to consume their electricity without paying for it. The electricity bill in this case should be charged directly to the consumer. Therefore, both parties need to have an agreement to charge the bill to the consumer as a service provided by the manager. Such practices are expected to grow with the increasingly widespread use of a Smart Micro-grid, which enables each consumer or organisation to generate and manage electricity usage locally. In the situation described above, the amount of consumption must be managed rigorously because it strongly affects the financial aspects of electricity consumption for both the consumer and the manager. In addition, technology to maintain the privacy of the consumer is also necessary because the manager may use this information in an indiscriminate manner in the future.

Achieving the features described above is difficult in the practical use of an application. In particular, the simplest way to keep the privacy of the consumer is for the manager of the outlets to charge the bill for electricity directly to the consumers on the spot. Such direct local payment gives the manager no information about the consumer except for their consumption. However, the management costs for this type of billing are high when there are a large number of consumers and heavy times of consumption. Although one might think that a lump payment of the total amount of consumption linked directly to the consumption information of the consumer can decrease management costs, such an approach can potentially leak the action history of the consumers to the manager. Simple anonymization such as replacement of an identifier for each consumer with a random value cannot guarantee the validity of consumption although the anonymity for the consumer can be guaranteed. Then, the manager might be at the risk of double spending or impersonation by other consumers. In other words, trade-offs exist between management costs, privacy, and security.

Metke and Ekl [6], for example, have discussed the necessity of cryptography for the Smart Grid, and He et al. [7] have concluded that the use of cryptography is a practical solution for the Smart Grid. Based on these works, we discuss the aforementioned problem via a cryptographic approach in our research. We also consider the high computational cost in the Smart Grid environment. Hence, we also need to discuss feasibility, i.e., the computational cost and system architecture of a cryptographic solution in addition to its construction.

This chapter is an extension of previous work from AINA Workshops 2017 [8]. In the previous work, we showed an anonymous authentication protocol with controllable-linkability group signatures and a new group signature scheme as its building block. In this work, we discuss the correctness and the security of the proposed group signature scheme as analysis of the scheme. We also implement the scheme on Raspberry Pi as a device simulating a smart meter in the C language to estimate the performance in the real world.

3.1.2 Contribution

In this work, we propose an anonymous authentication protocol to enable a consumer to anonymously utilise electrical power under a smart meter managed by a manager. Loosely speaking, whereas the privacy of a consumer can be guaranteed by the anonymity of the proposed protocol, the validity of electricity consumption can be guaranteed by its authentication mechanism. Furthermore, our anonymous authentication protocol prevents a consumer and a manager from denying the respective electricity consumption. In this way we can prevent potential risks such as double spending and billing fraud. We also propose a new signature scheme as a building block suitable for the proposed protocol. Moreover, we implement cryptographic parts of our protocol on Raspberry Pi to estimate the computational overhead. We then show that the computational time of the parts with our signature scheme can be finished within one-and-a-half seconds.

We describe the main idea of our anonymous authentication protocol below. To overcome the problems described in the previous section, we consider the linkability of consumers to a manager without downgrading anonymity for the consumers. We call such a feature *unlinkability* informally. (See Sect. 3.2.2 for details of unlinkability.) To this purpose, we first adopt *controllable-linkability group signatures (CL-GS)* [9] as the main building block. The anonymity for signers as standard group signatures [10] guarantees that only some group members can generate signatures accepted by a public key of the group and a verifier cannot identify the actual signer.

The CL-GS allows, in addition to the anonymity described above of the standard group signatures, an authorized entity called a link manager to identify whether group signatures are generated by the same signer or not without identifying the signer him/herself. By virtue of such CL-GS, our protocol is able to achieve the unlinkability of consumers against a manager for electricity usage without downgrading privacy. Likewise, the unforgeability and the traceability of CL-GS, whereby only a group manager can identify the actual signer, support the unforgeability and undeniability of charging information, respectively. Intuitively, unforgeability guarantees that no third party can generate valid group signatures whose verification is accepted. On the one hand, undeniability guarantees that a consumer cannot generate group signatures different from the ones generated in the past to deny the consumer's usage against a manager. See Sects. 3.2.2 and 3.3.1 for the details of these features.

Although our model and security analysis is informal, our protocol utilizes CL-GS as a black-box tool and hence is able to utilize any CL-GS. Performances of our protocol can be upgraded by adopting efficient CL-GS. In addition, as a potential instantiation of our protocol, we also propose a new framework of CL-GS called *token-dependent controllable-linkability group signatures (TDCL-GS)*, where signatures with the same token are linkable for a link manager. In other words, TDCL-GS allows a link manager to link a signer of signatures if both the tokens and the signer are the same. Since the intuition of TDCL-GS corresponds to the features of the proposed protocol described above by assigning a unique token to each smart meter, we believe that TDCL-GS is more suitable for our anonymous authentication protocol. See Sect. 3.4 for more details.

3.1.3 Related Work

3.1.3.1 Privacy Preservation on a Smart Grid

The closest works to ours with group signatures from the viewpoint of cryptographic constructions are the protocol by Diao et al. [5] and the protocol by Zargar and Yaghmaee [11]. However, note that these are still strictly different from our work since they deal only with electrical usage under a smart meter managed by the consumer him/herself. The main target of our work is privacy with electricity usage under a smart meter managed by a manager of outlets, i.e., not managed by a consumer. In addition, the protocol by Diao et al. has been broken by Qu et al. [12]. Alternatively, many protocols [13–19] with other cryptographic schemes such as homomorphic encryption to privately aggregate consumption of a smart meter exist, but there is no work dealing with a similar setting as ours. To the best of our knowledge, the same problem has been discussed in only our latest work [20]. We showed a protocol with ID-federation, where each consumer uniquely generates a pseudo-random identifier for each manager, to keep the privacy. However, the protocol forces consumers to manage pseudo-random identifiers depending on the number of managers and vice versa. Namely, the protocol might be difficult when a large number of participants exists. Finally, although there is the open smart grid protocol (OSGP) [21] as a communication protocol with privacy for the Smart Grid, Jovanovic et al. [22] have shown the vulnerabilities of OSGP.

3.1.3.2 Group Signatures

Group signatures [10] are digital signatures where members among a signing group can sign as a signer with anonymity. If any problem occurs, a group manager can open signatures to identify/revoke the actual signer. In group signatures, a class called linkable group signatures [23] allows a receiver to identify only whether the signatures are generated by the same signer or not. The first scheme by Nakanishi et al. [23] was public linking in which anyone could identify signatures generated by the same signer. Recently, Hwang et al. [9] have proposed CL-GS which allows only a link manager to identify whether the signatures are generated by the same signer or not. Several researchers [24, 25] have then followed in the area of CL-GS. In yet another approach, Emura et al. [26] have proposed a capability called time-token dependent linking where signatures with the same token are publicly linkable. Our instantiation is a natural extension of the scheme by Emura et al. to CL-GS.

3.2 Requirements

In this section, we define a system model of the Smart Grid. Next, we define the security requirements of the model.

3.2.1 Participants and Their Business Logic

In our model, the following participants exist:

Consumer A consumer is a person who uses electricity. We denote a consumer by $\mathscr{C}$.

Manager A manager is a person who manages outlets. Here, we define the outlets as interfaces for the use of electricity. We denote a manager by $\mathscr{M}$.

Electric Utility An electric utility is a utility which provides electricity. While it provides electricity to both the consumer and the manager, there are two types of electric utilities, i.e., one for the consumer and one for the manager. We denote an electric utility by $\mathscr{U}$, and utilities which provide electricity to $\mathscr{C}$ or $\mathscr{M}$, respectively by $\mathscr{U}_C$, $\mathscr{U}_M$.

Power Grid A power grid is an organization that provides an infrastructure for all the entities. Namely, it is a single entity and the other entities are connected via the power grid.

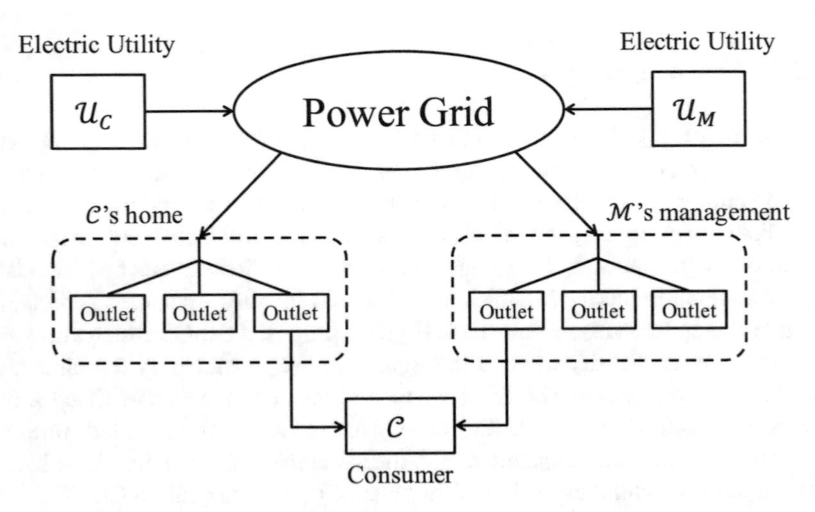

Fig. 3.1 Model of the smart grid in this work

The $\mathcal{C}$ has to pay the electric bill if $\mathcal{C}$ uses electricity in $\mathcal{C}$'s home. Similarly, $\mathcal{M}$ has to pay the electric bill if $\mathcal{M}$ uses electricity in $\mathcal{M}$'s home as a consumer. Suppose that $\mathcal{C}$ wants to use electricity outdoors as well as in $\mathcal{C}$'s home and $\mathcal{M}$ allows the consumer to use their electric resources.

When a consumer $\mathcal{C}$ starts to use electricity through an outlet under $\mathcal{M}$, $\mathcal{C}$ gives information about its electric utility $\mathcal{U}_C$ to $\mathcal{M}$. This information is stored in an IC card issued by $\mathcal{U}_C$. Then, $\mathcal{M}$ retrieves the electricity utilized by $\mathcal{C}$ from $\mathcal{U}_C$. Each record of the charging information by $\mathcal{C}$ at home and outside is consolidated to $\mathcal{U}_C$. $\mathcal{C}$ pays the electric bill only to $\mathcal{U}_C$.

$\mathcal{C}$ can utilize $\mathcal{M}$'s outlets by showing $\mathcal{C}$'s identifier given from $\mathcal{U}_C$. $\mathcal{M}$ sends the amount in charge for $\mathcal{C}$'s use of outlets. Then, $\mathcal{U}_C$ makes settlements for the amount in charge by $\mathcal{C}$. In this study, we deal with the flow of charging information when the consumer uses electricity outside in a Smart Grid. We also assume that the manager conforms oneself to the current specifications [1] for how the electricity is measured and the charge of the electricity. We show overview in Fig. 3.1.

3.2.2 Security Requirements

3.2.2.1 Assumption

A trustworthy utility is registered in the power grid and then is assigned an identifier ID_U from the power grid. We also assume that the utilities are trusted: more specifically, we assume that the utilities $\mathscr{U}_C$ and $\mathscr{U}_M$ do not collude with each other and do not fail in any operation such as the computation of the electricity bill. In general, these utilities are public domains authorized by a trusted entity such as government.

3.2.2.2 Unlinkability

An adversary of this requirement is a manager $\tilde{\mathscr{M}}$ who can interact with consumer independently of other managers. We assume that $\tilde{\mathscr{M}}$ does not collude with an honest manager $\mathscr{M}$ which the consumer connects with. Likewise, $\tilde{\mathscr{M}}$ does not collude with utility $\mathscr{U}_C$ since $\mathscr{U}_C$ always knows the charging information of the consumer. $\tilde{\mathscr{M}}$'s goal is to output a pair of the consumer's identifiers under an honest manager $\mathscr{M}$, who does not collude with $\tilde{\mathscr{M}}$, and under $\tilde{\mathscr{M}}$. We say a scheme is unlinkable if $\tilde{\mathscr{M}}$ cannot output the tuple of the consumer's identifiers under $\mathscr{M}$ and $\tilde{\mathscr{M}}$.

3.2.2.3 Undeniability

An adversary of this requirement is a malicious $\mathscr{C}$ who can interact with a manager. We say that $\mathscr{C}$ denies the charging information if $\mathscr{C}$ outputs different charging information, which is acceptable in $\mathscr{U}_C$, from that output by $\mathscr{C}$ in the past under interactions with $\mathscr{M}$. We say also that a scheme is undeniable for a consumer if $\mathscr{C}$ cannot deny the charging information. Likewise, we say that $\mathscr{M}$ denies the charging information if $\mathscr{M}$ outputs different charging information, which is acceptable in $\mathscr{U}_C$, from that output by $\mathscr{M}$ in the past under interactions with $\mathscr{C}$. $\mathscr{M}$'s goal is to deny the charging information, and we say that a scheme is undeniable for a manager if $\mathscr{M}$ cannot deny the charging information.

3.2.2.4 Unforgeability for Charging Information

This requirement is for the validity of the charging information of a consumer. The charging information must not be manipulated. An adversary of the requirement is all entities except for the consumer-self, and $\mathscr{U}_C$ and $\mathscr{U}_M$ can collude with each other only in the requirement. The consumer utilizes an outlet under its agreement and pays the electric bill. Namely, the main scenario is whether the other entities can generate any charging information of the consumer without the consumer's agreement or not.

3.3 Proposed Anonymous Authentication Protocol

We propose an authentication protocol for the Smart Grid. In this protocol, a manager authenticates a consumer anonymously when the consumer uses outlets owned by the manager. In other words, a consumer can keep privacy when the consumer utilizes outlets in any public place. First, we describe the algorithms of CL-GS and the digital signatures as the main building blocks. Next, we describe the ground rules and the construction of the proposed protocol.

3.3.1 Building Blocks

3.3.1.1 Controllable-Linking Group Signatures (CL-GS)

We recall the syntax of CL-GS and the security below [9].

Algorithm

CL-GS includes the following algorithms:

Setup takes as input a security parameter λ, and outputs a public parameter *params*.

GKeyGen takes as input *params*, and outputs a group public key gpk, a group master key gsk, a linking key lk, an initial revocation storage $\mathsf{grs} := \emptyset$ and an initial revocation list $\mathsf{RL}_0 := \emptyset$.

Join takes as input gsk,grs, a unique identity for signer ID and *params*, and outputs a signing key $\mathsf{sigk}_{\mathsf{ID}}$ and updated revocation storage grs.

GSign takes as inputs gpk, $\mathsf{sigk}_{\mathsf{ID}}$, a message m and *params*, and outputs group signature σ.

Revoke takes as inputs gpk, grs, a set of revoked users $\{\mathsf{ID}_1, \mathsf{ID}_2, \ldots, \mathsf{ID}_n\}$ and *params*, and outputs a revocation list RL_T.

GVerify takes as inputs gpk, RL_T, σ, m and *params*, and outputs *true* or *false*.

Link takes as inputs gpk, RL_T, two group signatures and messages (σ_0, m_0), (σ_1, m_1), lk and *params*, and outputs *true* if the two signatures are generated by the same signer, otherwise outputs false.

Open takes as inputs σ, grs, lk and *params*, and outputs ID of the member that generated these signatures.

Security Definition

Let CL-GS meet the following requirements. Although these requirements are informal, the formal definition has been shown in the existing work [9].

Correctness If a signer has not been revoked, GVerify outputs *true* on (gpk, σ, m) for its generated signatures on m by GSign. In addition, Link outputs *true* if signatures are generated by the same signer.

Anonymity A verifier of signatures cannot link actual signers of the signatures.

Unforgeability An adversary who has no signing key cannot generate signatures such that GVerify outputs *true*.

Linking Soundness If any two signatures are generated by different keys or for different tokens, Link output *false*.

Traceability If fraud occurs, Open is able to identify an actual signer of group signatures given as input.

3.3.1.2 Digital Signatures

Let (Gen, Sign, Verify) be a digital signature scheme. The key generation algorithm Gen takes as input a security parameter λ, and outputs a pair of verification/signing keys (vk, sigk). The signing algorithm Sign takes as input sigk and a message to be signed m, and outputs a signature Σ. The verification algorithm Verify takes as input vk, Σ and m, and outputs 0/1. We require the following correctness property: for all (vk, sigk) $\leftarrow$ Gen(1^λ) and m, $\Pr[\mathsf{Verify}(\mathsf{vk}, \mathsf{Sign}(\mathsf{sigk}, m), m) = 1] = 1$ holds. We say that a digital signature scheme is EUF-CMA if the probability, that $\mathsf{Verify}(\mathsf{vk}, \Sigma^*, m^*) = 1$ and an adversary did not send m^* as a signing query, is negligible.

3.3.2 Ground Rules of the Proposed Protocol

In this section, we define ground rules of the proposed protocol and its initial setting executed in advance.

3.3.2.1 Ground Rules for the Proposed Protocol

Our protocol consists of three phases: the Preparation Phase, Sign Phase, and Verification Phase. Each phase is executed as follows:

Preparation Phase The Preparation Phase is performed only once. In this phase, an electric utility $\mathscr{U}_C$ issues an identity ID and keys which are necessary for user authentication of a consumer $\mathscr{C}$.

Sign Phase The Sign Phase is performed every time $\mathscr{C}$ uses electricity under $\mathscr{M}$. In this phase, $\mathscr{M}$ and $\mathscr{C}$ generate signatures for the charging information of $\mathscr{C}$ measured by $\mathscr{M}$.

Verification Phase The Verification Phase is executed periodically, e.g., monthly. In this phase, $\mathscr{U}_C$ verifies that signatures are generated by $\mathscr{C}$ and $\mathscr{M}$.

3.3.2.2 Initial Setting

Group manager $\mathscr{U}_C$ has a public key gpk, a secret key gsk and a linking key lk for a group. Likewise, $\mathscr{M}$ is given an identifier by $\mathscr{U}_M$ and has a secret key sk and its corresponding public key pk for the group. $\mathscr{M}$ measures the amount for electricity when $\mathscr{C}$ uses electricity under a smart meter with an identifier T managed by $\mathscr{M}$. The correctness of these keys is guaranteed by PKI, and keys are available to everyone with certification. $\mathscr{U}_C$ manages a database which stores signatures generated by $\mathscr{C}$ and $\mathscr{M}$, charging information and its hash value h as one set, which updates every time $\mathscr{C}$ uses electricity. Note that we assume that Setup and $\mathsf{GKeyGen}$ have been executed already.

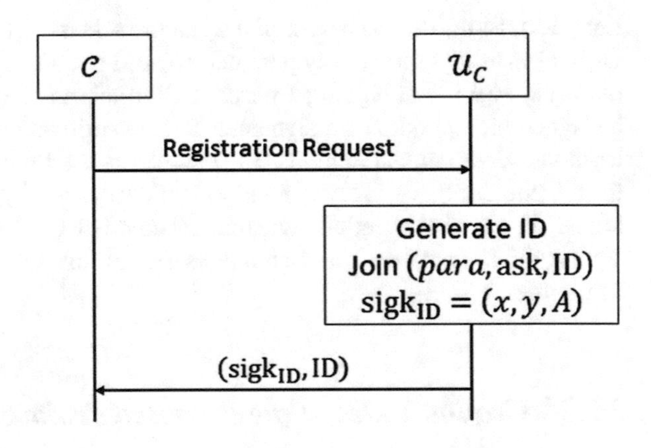

Fig. 3.2 The process of the preparation phase

3.3.3 *Construction*

In the proposed protocol, the charging information for each consumer is guaranteed by CL-GS. More specifically, each consumer $\mathscr{C}$ generates group signatures and each manager $\mathscr{M}$ must be able to know whether this $\mathscr{C}$ utilized $\mathscr{M}$'s outlets or not. Therefore, $\mathscr{M}$ can require $\mathscr{C}$ to pay its charging information via $\mathscr{U}_C$. The details of the construction are as follows:

3.3.3.1 Preparation Phase

The Preparation Phase is constructed as follows:

1. $\mathscr{C}$ sends a registration request to $\mathscr{U}_C$.
2. $\mathscr{U}_C$ generates an identity for $\mathscr{C}$ ID, and outputs $\mathsf{sigk}_{\mathsf{ID}} = (x, y, A)$. $\mathscr{U}_C$ then updates $\mathsf{grs} := \mathsf{grs} \cup \{(\mathsf{ID}, x)\}$ by executing $\mathsf{Join}(\mathsf{gsk}, \mathsf{grs}, \mathsf{ID}, params)$.
3. $\mathscr{U}_C$ sends $(\mathsf{sigk}_{\mathsf{ID}}, \mathsf{ID})$ to $\mathscr{C}$.

We show the process described above in Fig. 3.2.

3.3.3.2 Sign Phase

The Sign Phase is constructed as follows:

1. $\mathscr{M}$ generates (q, N, T), in which q is the amount of electricity in the charging information, N is a unique number, and T is an identifier of a smart meter. $\mathscr{M}$ generates a signature Σ for (q, N, T) by $\mathsf{Sign}(\mathsf{sk},(q, N,T))$. $\mathscr{M}$ sends a message $m = ((q, N, T), \Sigma)$ to $\mathscr{C}$.
2. $\mathscr{C}$ verifies the signature for q in m by $\mathsf{Verify}(\mathsf{pk}, \Sigma, m)$. $\mathscr{C}$ generates a signature $\sigma_C = \mathsf{GSign}(\mathsf{gpk}, \mathsf{sigk}_{\mathsf{ID}}, m, params)$ if Verify outputs *true*.
3. $\mathscr{C}$ sends (σ_C, m) to $\mathscr{U}_C$.
4. $\mathscr{U}_C$ extracts the newest $(\tilde{m}, \tilde{h})$ such that $\mathsf{Link}(\mathsf{gpk}, \mathsf{RL}_T, (\sigma_C, m), (\tilde{\sigma_C}, \tilde{m}), \mathsf{lk}, params)$ outputs *true* from the database. Then, $\mathscr{U}_C$ computes a hashed value $h = H(\tilde{h})$ and sends h to $\mathscr{C}$, where $\tilde{h}$ is the hashed value of the charging information for $\mathscr{C}$ under $\mathscr{M}$ in the past.

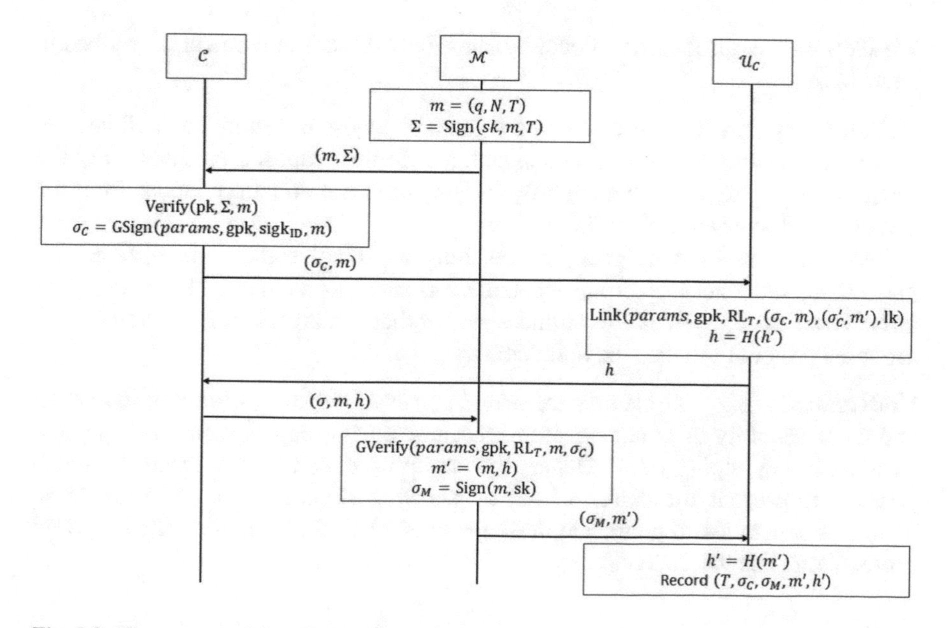

Fig. 3.3 The process of the signing phase

5. $\mathscr{C}$ sends (σ_C, m, h) to $\mathscr{M}$.
6. $\mathscr{M}$ verifies a signature σ_C by GVerify(gpk, RL_T, m, σ_C). $\mathscr{M}$ then sets $m' = m \parallel h$, and generates a signature $\sigma_M = Sign(m', \mathsf{sk})$ if GVerify outputs *true*. Otherwise, i.e., GVerify outputs *false*, $\mathscr{M}$ retries from step 1) again. Finally, $\mathscr{M}$ sends (σ_M, m') to $\mathscr{U}_C$.
7. $\mathscr{U}_C$ computes a hashed value $h' = H(m')$ of m', and adds $(T, \sigma_C, \sigma_M, m, h)$ as $(m, h) \leftarrow (m', h')$ to the database.

We show the process described above in Fig. 3.3.

3.3.3.3 Verification Phase

The Verification Phase is constructed as follows:

1. $\mathscr{U}_C$ extracts σ_C's for T and every signer from the database by Link.
2. $\mathscr{U}_C$ computes $h'_1 = H(m_1)$, $h'_i = H(m_i \parallel h'_{i-1})(2 \leq i \leq n)$. Then, $\mathscr{U}_C$ checks N_i in m_i to see whether N_i is used or not in the other messages m_i.
3. $\mathscr{U}_C$ verifies $\sigma_{C,n}$ and $\sigma_{M,n}$ by GVerify(gpk, $\mathsf{RL}_T, m_n, \sigma_{C,n}$) and Verify(pk, $\sigma_{M,n}$) if $h'_n = h_n$ holds.
4. $\mathscr{U}_C$ identifies an actual signer of $\sigma_{C,n}$ by Open and notifies q_i as the amount of electricity if GVerify and Verify output *true*. Otherwise, $\mathscr{U}_C$ tracks the falsification by verifying and opening all signatures.

3.3.4 *Security Analysis of the Proposed Protocol*

We show that our proposed protocol satisfies the security requirements described in Sect. 3.2.2.

Unlinkability No entity can link two signatures unless the entity has a liking key lk through the anonymity of group signatures. In the proposed protocol, only the group manager $\mathscr{U}_C$ has a linking key, and $\mathscr{U}_C$ does not collude with $\mathscr{M}$ from the assumption described in Sect. 3.2.2.

Moreover, the Link outputs *false* as long as actual signers are different by the linking soundness of group signatures. Hence, $\mathscr{M}$ cannot link the charging information of $\mathscr{C}$ even if $\mathscr{M}$ colludes with other managers $\mathscr{M}'$. Therefore, the proposed protocol satisfies the unlinkability.

Undeniability $\mathscr{U}_C$ can identify the actual signer of signatures through the Open and the traceability of group signatures, because each group signature is generated by a secret key $\mathsf{sigk}_{\mathsf{ID}}$ of $\mathscr{C}$. Hence, the denial of $\mathscr{C}$ can be prevented. It is also possible to prevent the denial of $\mathscr{M}$ by showing signatures of $\mathscr{M}$, because $\mathscr{C}$ generates signatures to guarantee those generated by $\mathscr{M}$. Therefore, the proposed protocol satisfies the undeniability.

Unforgeability The proposed protocol obviously satisfies the unforgeability since $\mathscr{C}$ generates signatures, which are unforgeable, for the charging information.

3.4 Instantiation: Token-Dependent Controllable-Linkability Group Signatures

In this section, we propose new group signatures called token-dependent controllable-linkability group signatures (TDCL-GS), where a manager who has a linking key can link only signatures with the same "token". By utilizing such a token as an identifier T of a smart meter, a more intuitive construction can be instantiated.

Although we leave as open problems to formally discuss its security, we can also utilize the existing CL-GS in the proposed protocol as described above.

3.4.1 *The Syntax of TDCL-GS*

TDCL-GS consists of the following algorithms:

Setup takes as input a security parameter λ, and outputs a public parameter *params*.

GKeyGen takes as input *params*, and outputs a group public key gpk, a group master key gsk, a linking key lk, an initial revocation storage $\mathsf{grs} := \emptyset$ and an initial revocation list $\mathsf{RL}_0 := \emptyset$.

TKeyGen takes as input *params*, and outputs a public key tpk and a secret key tsk.

Join takes as inputs gsk, grs, a unique identity ID for a signer and *params*, and outputs a signing key $\mathsf{sigk}_{\mathsf{ID}}$ and updated revocation storage grs.

TokenGen takes as inputs tsk, a identifier T and *params*, and outputs a token t_T.

GSign takes as inputs gpk, tpk, t_T, $\mathsf{sigk}_{\mathsf{ID}}$, a message m and *params*, and outputs a group signature σ.

Revoke takes as inputs gpk, grs, a set of revoked users for T $\{\mathsf{ID}_{T,1}, \mathsf{ID}_{T,2}, \ldots, \mathsf{ID}_{T,n}\}$ and *params*, and outputs a revocation list for T RL_T.

GVerify takes as inputs gpk, tpk, RL_T, σ, m and *params*, and outputs *true* or *false*. In some cases, this algorithm is the interactive algorithm of an entity that has a linking key.

Link takes as inputs gpk, tpk, RL_T, two group signatures and messages (σ_0, m_0), (σ_1, m_1), lk and *params*, and outputs *true* if the two signatures are generated by the same signer, otherwise outputs false.

Open takes as inputs σ, grs, lk and *params*, and outputs the ID of the member that generated these signatures.

Generally, group signatures have a Judge algorithm which verifies the validity of the output of Open. Emura et al. [26] omitted the Judge for efficiency. In this paper, we also omit the Judge from the proposed protocol.

3.4.2 Security Definition

In this section, we define the security requirements for TDCL-GS. The following requirements are informal, and a formal discussion is still an open problem.

Correctness If the signer has not been revoked, GVerify outputs true only if a correct signature is given and Link outputs *true* if some signatures are generated by the same signer.

Anonymity The entities cannot link signatures even if they have tsk.

Unforgeability An adversary who has no signing key cannot generate a signature such that GVerify outputs *true*.

Linking Soundness If the two signatures are generated by different keys or for different tokens, Link does not output *true*.

Traceability If fraud occurs, Open is able to identify the real signer of the group signature given as input.

3.4.3 Construction

In the proposed scheme, we use a unique identifier of a smart meter as a token T. A link manager who has a linking key can link signatures if these signatures are generated by the same signer using the same token. Hence, we add a linking key to GS-TDL and define the link manager as an entity who has a linking key. Hereinafter, we denote by the left-side arrow the random generation of each parameter. Also, let $\mathcal{G}$ be a probabilistic polynomial-time algorithm that takes the security parameter λ as input and generates the parameter $(p, \mathbb{G}_1, \mathbb{G}_2, \mathbb{G}_T, e, g_1, g_2)$ of bilinear groups, where p is a λ-bit prime, $\mathbb{G}_1$, $\mathbb{G}_2$ and $\mathbb{G}_T$ are groups of order p, e is a bilinear map from $\mathbb{G}_1 \times \mathbb{G}_2$ to $\mathbb{G}_T$, and g_1 and g_2 are generators of $\mathbb{G}_1$ and $\mathbb{G}_2$, respectively. Here we use the asymmetric setting.

Setup(1^λ): Let $(\mathbb{G}_1, \mathbb{G}_2, \mathbb{G}_3)$ be a bilinear group with prime order p, where $\langle g \rangle = \mathbb{G}_1$, $\langle g \rangle = \mathbb{G}_2$ and $e : \mathbb{G}_1 \times \mathbb{G}_2 \rightarrow \mathbb{G}_T$ be a bilinear map. Output $params = (\mathbb{G}_1, \mathbb{G}_2, \mathbb{G}_T, e, g_1, g_2, (\mathsf{Gen}, \mathsf{Sign}, \mathsf{Verify}))$.
GKeyGen$(params)$: Choose $(\gamma, l) \leftarrow \mathbb{Z}_p$, $(h, u) \leftarrow \mathbb{G}_1^2$, $k \leftarrow \mathbb{G}_2$, and compute $W = g_2^\gamma$, $f \leftarrow u^l$.
TKeyGen$(params)$: Output token key $(\mathsf{tpk}, \mathsf{tsk}) \leftarrow \mathsf{Gen}(1^\lambda)$.

Join(gsk, grs, ID, *params*): Choose $(x, y) \leftarrow \mathbb{Z}_p$, compute $A = (g_1 h^{-y})^{\frac{1}{y+x}}$, output $\mathsf{sigk}_{\mathsf{ID}} = (x, y, A)$, and update grs := grs $\cup$ {ID, x}.
TokenGen(tsk, T, *params*): Let the token be $T \in \mathbb{Z}_p$. Compute $W_T = g_2^T$ and $\Sigma \leftarrow$ Sign(tsk, W_T), and output $t_T = (T, W_T, \Sigma)$.
GSign(gpk, tpk, t_T, $\mathsf{sigk}_{\mathsf{ID}}$, m, *params*): If Verify(tpk, W_T, Σ)=*false*, then output *false*. Otherwise, choose $(\alpha, \beta) \leftarrow \mathbb{Z}_p^2$, set $\delta = \beta x - y$, and compute $C = Ah^{\beta}$, $\tau = g_1^{\frac{1}{x+T}} f^{\frac{\alpha}{x+T}}$ and $\tau' = u^{\frac{\alpha}{x+T}}$. Choose $(r_x, r_\delta, r_\beta) \leftarrow \mathbb{Z}_p^3$, and compute as follows:

$$R_1 = \frac{e(h, g_2)^{r_\delta} e(h, W)^{r_\beta}}{e(C, g_2)^{r_x}}, \quad R_2 = e(\tau, g_2)^{r_x}, \quad c = H(\mathsf{gpk}, \mathsf{tpk}, C, \tau, \tau', R_1, R_2, m),$$

$$s_x = r_x + cx, \quad s_\delta = r_\delta + c\delta, \quad s_\beta = r_\beta + c\beta.$$

Output $\sigma = (C, \tau, \tau', c, s_x, s_\delta, s_\beta)$.
Revoke(gpk, grs, $\{\mathsf{ID}_{T,1}, \mathsf{ID}_{T,2}, \ldots, \mathsf{ID}_{T,n}\}$, *params*): If there exists ID $\in \{\mathsf{ID}_{T,1}, \ldots, \mathsf{ID}_{T,n}\}$ that is not joined to the system via the Join, then output *false*. Otherwise, extract $(\mathsf{ID}_{T,1}, x_{T,1}), \ldots, (\mathsf{ID}_{T,n}, x_{T,n})$ from grs. Output $\mathsf{RL}_T :=$ $\{(\mathsf{ID}_{T,1}, e(g_1^{\frac{1}{x_{T,1}+T}}, k)), \ldots, (\mathsf{ID}_{T,n_T}, e(g_1^{\frac{1}{x_{T,n_T}+T}}, k))\}$.
GVerify(gpk, tpk, RL_T, σ, m, *params*): Assume that Verify(tpk, W_T, Σ) = *true* (if not, output *false*). Inquire whether (τ, τ') of $\sigma = (C, \tau, \tau', c, s_x, s_\delta, s_\beta)$ are included in RL_T to group manager.[1] If (τ, τ') are contained in RL_T, then output *false*. Otherwise, compute

$$R_1{}' = \frac{e(h, g_2)^{s_\delta} e(h, W)^{s_\beta}}{e(C, g_2)^{s_x}} \left(\frac{e(C, W)}{e(g_1, g_2)}\right)^{-c}, \quad R_2{}' = e(\tau, g_2)^{s_x} \frac{e(g_1, g_2) e(f^{\alpha}, g_2)}{e(\tau, W_T)}^{-c},$$

and output *true* if $c = H(\mathsf{gpk}, \mathsf{tpk}, C, \tau, \tau', R_1', R_2', m)$ holds, and *false* otherwise.
Link(gpk, tpk, RL_T, (σ_0, m_0), (σ_1, m_1), lk, *params*): Given two message (m_0, m_1) and their corresponding signatures (σ_0, σ_1), and linking key lk, this algorithm tries to find links among signatures if they are generated from the same signer i. It first verifies the signatures' validity by using GVerify. Then, it checks $e(\tau_0/\tau_1, k) \stackrel{?}{=} e(\tau_0'/\tau_1', k^l)$. It returns *true* if successful, otherwise outputs *false*.
Open(σ, grs, lk, *params*): Choose x from grs, and output the ID corresponding to x where $(\frac{e(\tau,k)}{e(\tau',k^l)})^{x+T} = e(g_1, k)$ by using σ and linking key lk.

3.4.4 *Security Analysis*

We discuss whether the proposed protocol meets the security requirements. First, we recall several assumptions.

[1] If these signatures are not required for checking, the verifier of GVerify can compute without interaction to the group manager.

3.4.4.1 Assumptions

In the following scheme, we assume the simultaneous decisional Diffie-Hellman inversion (SDDHI) assumption [27] and the q-strong Diffie-Hellman (SDH) assumption [28]. We recall the definitions below.

SDDHI Assumption: We say that the SDDHI assumption holds if for all PPT adversaries $\mathscr{A}$,(Pr[$(p, \mathbb{G}_1, \mathbb{G}_2, \mathbb{G}_\gamma, e, g_1, g_2) \leftarrow \mathscr{G}(1^\lambda); x \leftarrow \mathbb{Z}_p; (\gamma, st) \leftarrow \mathscr{A}^{\mathscr{O}_x}(p, \mathbb{G}_1, \mathbb{G}_2, \mathbb{G}_\gamma, e, g_1, g_2, g_1^x); \tau_0 = g_1^{\frac{1}{x+\gamma}}; \tau_1 \leftarrow \mathbb{G}_1; b \leftarrow \{0,1\}; b' \leftarrow \mathscr{A}^{\mathscr{O}_x}(y_b, st) : b = b'$]$-\frac{1}{2}$) is negligible, where $\mathscr{O}_x$ is an oracle, which takes as input $z \in \mathbb{Z}_p^* \backslash \{T\}$ and outputs $g_1^{\frac{1}{x+z}}$.

q-SDH Assumption: We say that the q-SDH assumption holds if for all PPT adversaries $\mathscr{A}$, (Pr[$(p, \mathbb{G}_1, \mathbb{G}_2, \mathbb{G}_T, e, g_1, g_2) \leftarrow \mathscr{G}(1^\lambda); \gamma \leftarrow \mathbb{Z}_p; (x, g_1^{\frac{1}{x+\gamma}}) \leftarrow \mathscr{A}(p, \mathbb{G}_1, \mathbb{G}_2, \mathbb{G}_T, e, g_1, g_1^\gamma, \ldots, g_1^{\gamma^q}, g_2, g_2^\gamma); x \in \mathbb{Z}_p^* \backslash \{-\gamma\}$]) is negligible.

3.4.4.2 Correctness

We show a calculation process for the GVerify algorithm in the Eqs. (3.1) and (3.2). Similarly, we show the calculation process for Link in the Eqs. (3.3) and (3.4), where $T = T'$ and $x = x'$ hold for these equations if σ and σ' are generated by the same signer under the same smart meter.

$$
\begin{aligned}
R_1' &= \frac{e(h, g_2)^{s_\delta} e(h, W)^{s_\beta}}{e(C, g_2)^{s_x}} \cdot \left(\frac{e((g_1 h^{-y})^{\frac{1}{\gamma+x}} h^\beta, g_2^\gamma)}{e(g_1, g_2)} \right)^{-c} \\
&= \frac{e(h, g_2)^{s_\delta} e(h, W)^{s_\beta}}{e(C, g_2)^{s_x}} \cdot \left(\frac{e((g_1 h^{-y}), g_2)^{\frac{\gamma}{\gamma+x}} e(h, g_2^\gamma)^\beta}{e(g_1, g_2)} \right)^{-c} \\
&= \frac{e(h, g_2)^{r_\delta + c(\beta x - y)} e(h, W)^{r_\beta} e(g_1, g_2)^{\frac{-c\gamma}{\gamma+x}} e(h^{-y}, g_2)^{\frac{-c\gamma}{\gamma+x}}}{e(C, g_2)^{r_x + cx} e(g_1, g_2)^{-c \cdot \frac{\gamma+x}{\gamma+x}}} \\
&= \frac{e(h, W)^{r_\beta}}{e(C, g_2)^{r_x + cx}} \cdot \frac{e(h, g_2)^{r_\delta + c(\beta x - y)} e(h^{-y}, g_2)^{\frac{-c\gamma}{\gamma+x}}}{e(g_1, g_2)^{\frac{-cx}{\gamma+x}}} \\
&= \frac{e(h, W)^{r_\beta}}{e(C, g_2)^{r_x + cx}} \cdot \frac{e(h, g_2)^{r_\delta + c(\beta x - y)}}{e(g_1, g_2)^{-\frac{cx}{\gamma+x}} e(h^{-y}, g_2)^{\frac{c\gamma}{\gamma+x}}} \\
&= \frac{e(h, W)^{r_\beta} e(h, g_2)^{r_\delta} e(h, g_2)^{c\beta x} e(h, g_2)^{-cy}}{e(C, g_2)^{r_x} e(C, g_2)^{cx} e(g_1, g_2)^{-\frac{cx}{\gamma+x}} e(h^{-y}, g_2)^{\frac{c\gamma}{\gamma+x}}}
\end{aligned}
$$

$$= \frac{e(h, W)^{r_\beta} e(h, g_2)^{r_\delta}}{e(C, g_2)^{r_x}} \cdot \frac{e(h, g_2)^{c\beta x} e(h, g_2)^{-cy}}{e((g_1 h^{-y})^{\frac{1}{\gamma+x}} h^\beta, g_2)^{cx} e(g_1^{\frac{1}{\gamma+x}}, g_2)^{-cx} e(h^{\frac{-y}{\gamma+x}}, g_2)^{c\gamma}}$$

$$= \frac{e(h, W)^{r_\beta} e(h, g_2)^{r_\delta}}{e(C, g_2)^{r_x}} \cdot \frac{e(h, g_2)^{-cy}}{e(h, g_2)^{\frac{-cxy}{\gamma+x}} e(h, g_2)^{\frac{-cy\gamma}{\gamma+x}}} = R_1. \quad (3.1)$$

$$\begin{aligned} R_2{}' &= e(\tau, g_2)^{s_x} \cdot \left(\frac{e(g_1, g_2) e(f^\alpha, g_2)}{e(\tau, W_T)} \right)^{-c} \\ &= e(\tau, g_2)^{s_x} \cdot e(g_1, g_2)^{-c} e(f^\alpha, g_2)^{-c} e(g_1^{\frac{1}{x+T}} f^{\frac{\alpha}{x+T}}, g_2^T)^c \\ &= e(\tau, g_2)^{s_x} \cdot e(g_1, g_2)^{\frac{-cx-cT+cT}{x+T}} e(f^\alpha, g_2)^{\frac{-cx-cT+cT}{x+T}} \\ &= e(\tau, g_2)^{s_x} \cdot e(g_1^{\frac{1}{x+T}}, g_2)^{-cx} e(f^{\frac{\alpha}{x+T}}, g_2)^{-cx} \\ &= e(\tau, g_2)^{r_x+cx} \cdot e(\tau, g_2)^{-cx} = e(\tau, g_2)^{r_x} = R_2 \end{aligned} \quad (3.2)$$

$$\begin{aligned} e\left(\frac{\tau_1}{\tau_1'}, k\right) &= e\left(\frac{g_1^{\frac{1}{x+T}} f^{\frac{\alpha}{x+T}}}{g_1^{\frac{1}{x'+T'}} f^{\frac{\alpha'}{x'+T'}}}, k\right) = e\left(\frac{f^{\frac{\alpha}{x+T}}}{f^{\frac{\alpha'}{x'+T'}}}, k\right) \\ &= e\left(\frac{u^{\frac{\alpha}{x+T}}}{u^{\frac{\alpha'}{x'+T'}}}, k\right)^l \end{aligned} \quad (3.3)$$

$$e\left(\frac{\tau_2}{\tau_2'}, k^l\right) = e\left(\frac{u^{\frac{\alpha}{x+T}}}{u^{\frac{\alpha'}{x'+T'}}}, k^l\right) = e\left(\frac{u^{\frac{\alpha}{x+T}}}{u^{\frac{\alpha'}{x'+T'}}}, k\right)^l \quad (3.4)$$

Since the above equations hold, and the GVerify and Link algorithms output true, our scheme achieves the correctness.

3.4.4.3 Anonymity

Let the secret keys of any two signers be (x, x') and their resultant signatures be (σ, σ'), respectively. Each signature can then be written as $\tau = g_1^{\frac{1}{x+T}} f^{\frac{\alpha}{x+T}} = \tilde{\omega}^{\frac{1}{x+T}}$ and $\tau' = g_1^{\frac{1}{x'+T}} f^{\frac{\alpha}{x'+T}} = \tilde{\omega}^{\frac{1}{x'+T}}$, where we let each α be a constant number for the sake of convenience. Then, $\tilde{\omega}$ is identical to τ_0 or τ_1 of the true SDDHI tuple described in the SDDHI assumption. In particular, x' is random from the view of x, and hence τ' can be viewed as random from a standpoint of τ. That is, distribution of (σ, σ') is identical to that of the SDDHI assumption. If an adversary can defeat the anonymity by distinguishing the origin of the secret key from the given signatures, it implicitly means that the distribution of the SDDHI assumption can be distinguished. Since this contradicts the SDDHI assumption, the proposed scheme achieves the anonymity.

3.4.4.4 Unforgeability

Unforgeability can be regarded as $\tau = g_1^{\frac{1}{x+T}} f^{\frac{\alpha}{x+T}} = \tilde{g_1}^{\frac{1}{x+T}}$ if we assume that $\tilde{g_1} = g_1 f^{\alpha}$ in τ and τ' are generated by the GSign. Likewise, unforgeability can be regarded as $\tau' = u^{\frac{\alpha}{x+T}} = \tilde{u}^{\frac{1}{x+T}}$ if we assume $\tilde{u} = u^{\alpha}$.

This can be regarded as q-SDH assumption to compute $g_1^{\frac{1}{x+T}}$ for any T. Then, the unforgeability can be reduced to the q-SDH assumption.

3.4.4.5 Linking Soundness

For any two signatures, the Eqs. (3.3) and (3.4) are equal if $x_0 + T_0 = x_1 + T_1$, $\alpha_0 = \frac{x_0+T_0}{x_1+T_1}$ or $\alpha_1 = \frac{x_1+T_1}{x_0+T_0}$ holds. Moreover, Link does not accept signatures signed by different signers since the probability that the equations in the Link hold is $\frac{1}{p}$, where p is an order of a group. That is, the probability of the event is negligibly small. Our proposed scheme thus achieves the linking soundness.

3.4.4.6 Traceability

A signer is uniquely determined from a public parameter if valid signatures are given as input for Open. We show a calculation process for Open as in the Eq. (3.5).

$$\left(\frac{e(\tau_1, k)}{e(\tau_2, k^l)}\right)^{x+T} = \left(\frac{e(g_1^{\frac{1}{x+T}} f^{\frac{\alpha}{x+T}}, k)}{e(u^{\frac{\alpha}{x+T}}, k)^l}\right)^{x+T} = \left(\frac{e(g_1^{\frac{1}{x+T}}, k)e(u^{\frac{\alpha}{x+T}}, k)^l}{e(u^{\frac{\alpha}{x+T}}, k)^l}\right)^{x+T}$$
$$= e(g_1^{\frac{1}{x+T}}, k)^{x+T} = e(g_1, k) \tag{3.5}$$

Thus, our scheme achieves the traceability.

3.4.5 *Application to the Proposed Protocol*

The TDCL-GS described above is applicable to the proposed protocol by substituting each algorithm for that of the CL-GS in Sect. 3.3.3. More specifically, each $\mathcal{M}$

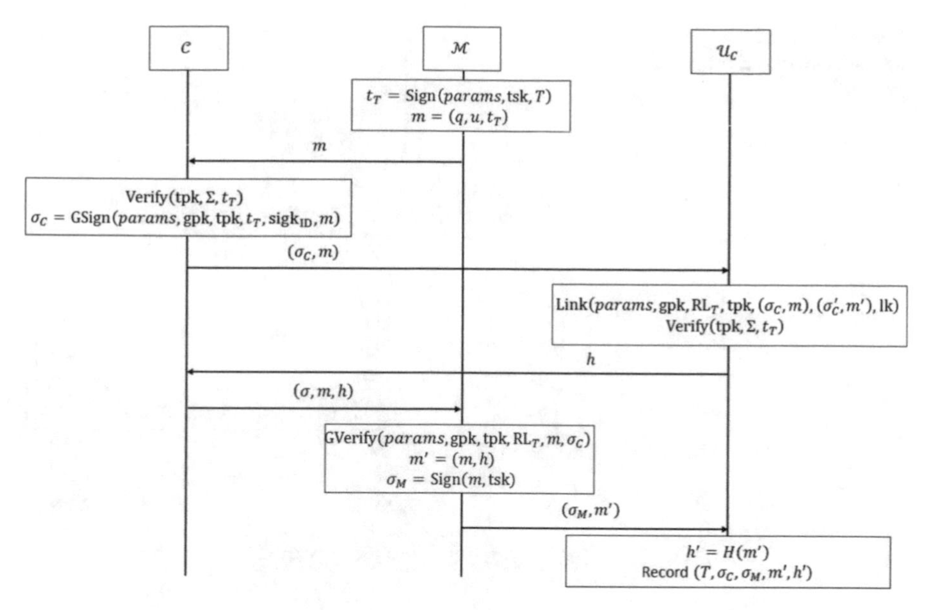

Fig. 3.4 Process of signing phase with TDCL-GS

owns (tsk, tpk) and generates t_T by executing the TokenGen. The token is utilized together with an identifier T for any smart meter.

We omit the details of the construction, but show the process of the Sign Phase with TDCL-GS in Fig. 3.4.

3.5 Discussion

In this section, we discuss the feasibility of the proposed protocol. First, we discuss the potential architecture of the proposed protocol in the real world although its social implementation is still an open problem. Next, we show the performance of the proposed protocol via implementation of the cryptographic parts including the scheme shown in the previous section.

3.5.1 *The Potential Architecture of the Proposed Protocol*

In this section, we discuss the architecture in which the proposed anonymous authentication protocol is utilized. First, we suggest that the proposed protocol is executed between a smart meter managed by a manager and a smart-phone owned by a consumer when the consumer utilizes outlets owned by the manager. More specifically, the smart-phone communicates with the outlets via contactless communication, and then the outlets forward their topics to the smart meter

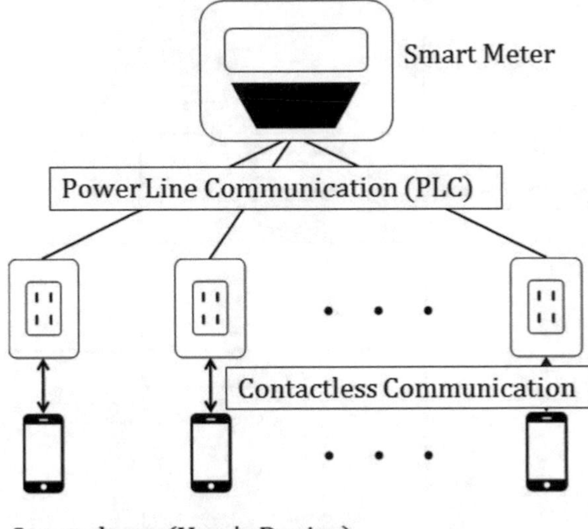

Fig. 3.5 Potential architecture of our proposed protocol

via the power line communication (PLC). We note that there are authentication-based outlets[2] where the outlets are activated by contactless communication via IC cards. The above communication method can be instantiated by the use of the authentication-based outlets. We show the construction of the architecture in Fig. 3.5.

Here, we also note that the computation steps related to the cryptographic schemes are run only by the smart-phone and the smart meter, and not the outlets. Since cryptographic computation is heavy in general, it should be run by a computationally rich device. Although a smart-phone is sufficient for such computation, unfortunately, to the best of our knowledge, there is no smart meter with such computational power. Therefore, we use Raspberry Pi[3] instead of the existing smart meters. Raspberry Pi is a single-board computer controlled by an ARM microcomputer, which is used for many smart meters. The Raspberry Pi has the computational power equivalent to a smart phone and so can compute cryptographic schemes. In fact, much research [29–34] utilizes the Raspberry Pi in implementing their proposed protocols instead of a smart meter.

Our protocol also requires a public key infrastructure (PKI) to manage public keys because our protocol utilizes public key cryptography. We suggest the establishment of a public domain, e.g., such as a government or an administrative agency, as the entity to operate the PKI. For example, one organization Japan, the Organization for Cross-regional Coordination of Transmission Operators

[2]Sony Japan, "Authentication-Based Outlets". (In Japanese.) https://www.sony.co.jp/SonyInfo/News/Press/201202/12-023/.

[3]https://www.raspberrypi.org.

Table 3.1 Experimental environment

	Mac Pro	Raspberry Pi
CPU	Intel Core i5 @2.7GHz	ARM Cortex-A53 @1.2GHz
Memory size	8GB	1GB
OS	Mac OS 10.11.2	Raspbian GNU/Linux 8.0
Compiler	gcc Apple LLVM version 8.0.0	gcc 4.9.2
Library	TEPLA 2.0	TEPLA 2.0

Table 3.2 Computational time for each step [sec]

	Mac Pro	Raspberry Pi
GSign	0.063	0.257
GVerify	0.147	0.586
Link	0.014	0.060
Open	0.715	2.915

(OCCTO)[4] has been established to realize a stable electricity supply in the Smart Grid. Currently, each company has to register its service information in OCCTO in order to expand a new service on the Smart Grid. Hence, we assume the existence of a trusted organization such as OCCTO in each region, and also assume that such an organization provides the PKI to utilize our protocol. Furthermore, we assume that a secret key and its public key are registered in each device in advance.

3.5.2 *Experiment*

In this section, we implement the cryptographic parts of the proposed protocol along with the environment described in the previous section.

3.5.2.1 Experimental Result

We show the computational time for each step in Table 3.2, and the results with respect to the number of simultaneous connections on the Raspberry Pi in Figs. 3.6 and 3.7. According to Table 3.2, the computational time for the proposed protocol can be estimated within one-and-a-half seconds because the computation steps on the Raspberry Pi are a single execution of GSign, GVerify, and of a standard signature scheme such as ECDSA.

Figures 3.6 and 3.7 show that parallelization is useful for decreasing the effect of simultaneous connections. The advantage of parallelization fully depends on the number of cores of a device. In particular, Raspberry Pi in this experiment owns

[4]https://www.occto.or.jp.

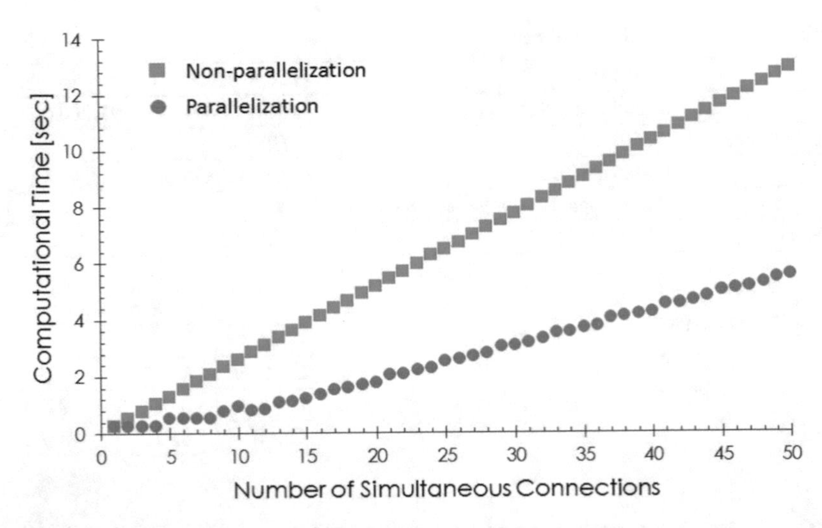

Fig. 3.6 The computational time of GSign on Raspberry Pi

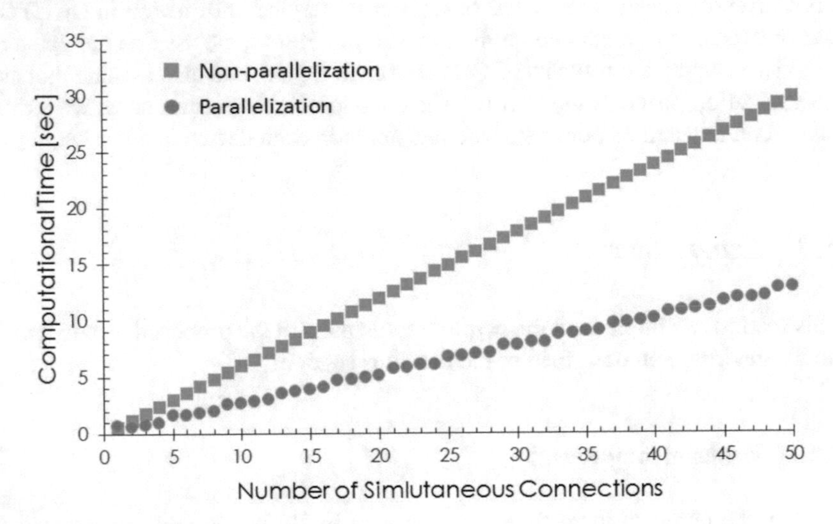

Fig. 3.7 The computational time of GVerify on Raspberry Pi

four cores, and so the computational time increases per four connections. This time increases linearly with respect to the number of connections, and hence we can estimate the computational time for any number of connections. For instance, we need about 10 s for GSign and about 25 s for GVerify to deal with 100 connections. Therefore, the total computational time can be estimated as about 40 s for 100 connections in the environment described above. We also note that the computational time can be decreased by increasing the number of cores as described above. Namely, we need to set the number of devices along with the possible number of simultaneous connections.

3.6 Conclusion

In this paper, we have focused on future services whereby a consumer pays an electricity bill to utilize outlets in public places over the Smart Grid. In addition, we have and proposed an anonymous authentication protocol which is based on a group signature scheme with controllable linkability. We have also proposed a new framework for controllable-linkability group signatures with tokens, called token-dependent controllable-linkability group signatures (TDCL-GS), as an instantiation suitable for the proposed protocol. Although the security analysis of TDCL-GS is informal, we believe that such a scheme is more practical for our protocol in the Smart Grid. We plan to formalize and prove the security of both the proposed protocol and TDCL-GS in future work. We also plan to execute experiments that include communication overhead between devices to estimate the performance of our proposed protocol.

Acknowledgements We would like to thank professor Toru Fujiwara at Osaka University and Keita Emura at NICT for their valuable comments. We would also like to thank JSPS KAKENHI grant number 16K16065, and Secom Science and Technology Foundation.

References

1. *NIST framework and roadmap for smart grid interoperability standards, release 1.0.* National Institute of Standards and Technology, 2010.
2. W. Wang and Z. Lu, "Cyber security in the smart grid: Survey and challenges," *Computer Networks*, vol. 57, no. 5, pp. 1344–1371, 2013.
3. Z. Erkin and T. Veugen, "Privacy enhanced personal services for smart grids," in *Proc. of SEGS 2014*. ACM, 2014, pp. 7–12.
4. H. Y. Lam, G. S. K. Fung, and W. K. Lee, "A novel method to construct taxonomy electrical appliances based on load signatures," *IEEE Transactions on Consumer Electronics*, vol. 53, no. 2, pp. 653–660, 2007.
5. F. Diao, F. Zhang, and X. Cheng, "A privacy-preserving smart metering scheme using linkable anonymous credential," *IEEE Transactions on Smart Grid*, vol. 6, no. 1, pp. 461–467, 2015.
6. A. R. Metke and R. L. Ekl, "Security technology for smart grid networks," *IEEE Transactions on Smart Grid*, vol. 1, no. 1, pp. 99–107, 2010.
7. D. He, S. Chan, Y. Zhang, M. Guizani, C. Chen, and J. Bu, "An enhanced public key infrastructure to secure smart grid wireless communication networks," *IEEE Network*, vol. 28, no. 1, pp. 10–16, 2014.
8. H. Kishimoto, N. Yanai, and S. Okamura, "An anonymous authentication protocol for smart grid," in *Proc. of AINA Workshop 2017*. IEEE, 2017, pp. 62–67.
9. J. Y. Hwang, S. Lee, B. H. Chung, H. S. Cho, and D. Nyang, "Short group signatures with controllable linkability," in *Proc. of LightSec 2011*. IEEE, 2011, pp. 44–52.
10. D. Chaum and E. Van Heyst, "Group signatures," in *Proc. of EUROCRYPT 1991*, ser. LNCS, vol. 547. Springer, 1991, pp. 257–265.
11. S. Hajy, M. Zargar, and M. H. Yaghmaee, "Privacy preserving via group signature in smart grid," in *Proc. of EIAC 2013*, 2013.
12. H. Qu, P. Shang, X.-J. Lin, and L. Sun, "Cryptanalysis of a privacy-preserving smart metering scheme using linkable anonymous credential," Cryptology ePrint Archive, Report 2015/1066, 2015.

13. R. Bobba, H. Khurana, M. AlTurki, and F. Ashraf, "Pbes: A policy based encryption system with application to data sharing in the power grid," in *Proc. of ASIACCS 2009*. ACM, 2009, pp. 262–275.
14. M. Jawurek, M. Johns, and F. Kerschbaum, "Plug-in privacy for smart metering billing," in *Proc. of PETS 2011*, ser. LNCS, vol. 6794. Springer, 2011, pp. 192–210.
15. T. Dimitriou and G. Karame, "Privacy-friendly planning of energy distribution in smart grids," in *Proc. of SEGS 2014*. ACM, 2014, pp. 1–6.
16. F. D. Garcia and B. Jacobs, "Privacy-friendly energy-metering via homomorphic encryption," in *Proc. of STM 2010*, ser. LNCS, vol. 6710. Springer, 2011, pp. 226–238.
17. K. Kursawe, G. Danezis, and M. Kohlweiss, "Privacy-friendly aggregation for the smart-grid," in *Proc. of PETS 2011*, ser. LNCS, vol. 6794. Springer, 2011, pp. 175–191.
18. A. Rial and G. Danezis, "Privacy-preserving smart metering," in *Proc. of PETS 2011*. ACM, 2011, pp. 49–60.
19. X.-F. Wang, Y. Mu, and R.-M. Chen, "An efficient privacy-preserving aggregation and billing protocol for smart grid," *Security and Communication Networks*, vol. 9, pp. 4536–4547, 2016.
20. H. Kishimoto, N. Yanai, and S. Okamura, "Spacis: Secure payment protocol for charging information over smart grid," *Journal of Information Processing*, vol. 25, no. 1, pp. 12–21, 2017.
21. ETSI, "Open smart grid protocol (osgp)," Refference DGS/OSG-001, European Telecommunications Standards Institute, Sophia Antipolis Cedex, January 2012.
22. P. Jovanovic and S. Neves, "Practical cryptanalysis of the open smart grid protocol," in *Proc. of FSE 2015*, ser. LNCS, vol. 9054. Springer, 2015, pp. 297–316.
23. T. Nakanishi, T. Fujiwara, and H. Watanabe, "A linkable group signature and its application to secret voting," *Transactions of Information Processing Society of Japan*, vol. 40, no. 7, pp. 3085–3096, 1999.
24. J. Y. Hwang, S. Lee, B.-H. Chung, H. S. Cho, and D. Nyang, "Group signatures with controllable linkability for dynamic membership," *Information Sciences*, vol. 222, pp. 761–778, 2013.
25. M. S. I. Mamun and A. Miyaji, "Secure vanet applications with a refined group signature," in *Proc. of PST 2014*. IEEE, 2014, pp. 199–206.
26. K. Emura and T. Hayashi, "A light-weight group signature scheme with time-token dependent linking," in *Proc. of LightSec 2015*, ser. LNCS, vol. 9542. Springer, 2016, pp. 37–57.
27. J. Camenisch, S. Hohenberger, M. Kohlweiss, A. Lysyanskaya, and M. Mcycrovich, "How to win the clonewars: Efficient periodic n-times anonymous authentication," in *Proc. of ACM CCS 2006*. ACM, 2006, pp. 201–210.
28. D. Boneh and X. Boyen, "Short signatures without random oracles and the SDH assumption in bilinear groups," *Journal of Cryptology*, vol. 21, no. 2, pp. 149–177, 2008.
29. J. Hastings, D. Laverty, and D. J. Morrow, "A smart grid information system for demand side participation: Remote control of domestic appliances to balance demand," in *Proc. of UPEC 2013*. IEEE, 2013, pp. 1–5.
30. J. Wilcox, D. Kaleshi, and M. Sooriyabandara, "Director: A distributed communication transport manager for the smart grid," in *Proc. of ICC 2014*. IEEE, 2014, pp. 4227–4232.
31. S. Uludag, K.-S. Lui, W. Ren, and K. Nahrstedt, "Practical and secure machine-to-machine data collection protocol in smart grid," in *Proc. of CNS 2014*. IEEE, 2014, pp. 85–90.
32. M. Armendariz, M. Chenine, L. Nordstrom, and A. Al-Hammouri, "A co-simulation platform for medium/low voltage monitoring and control applications," in *Proc. of ISGT 2014*. IEEE, 2014, pp. 1–5.
33. H. Li, G. Dan, and K. Nahrstedt, "Lynx: Authenticated anonymous real-time reporting of electric vehicle information," in *Proc. of SmartGridComm 2015*. IEEE, 2015, pp. 599–604.
34. D. Smith, C. Olariu, P. Perry, and J. Murphy, "Impact of non-deterministic software execution times in smartgrid applications," in *Proc. of ISSC 2015*. IEEE, 2015, pp. 1–5.

Chapter 4
Attacks on Authentication and Authorization Models in Smart Grid

Trupil Limbasiya and Aakriti Arya

Abstract The evolution of a conventional electric grid infrastructure can be dated back to 1880s when the outstanding sources of energy were based on hydraulics and gas energy. But one cannot only depend upon the classic electric grid system in today's digital world. However, the smart grid and smart micro grid provide electric power in many efficient and measured ways, which are helpful in the technology-enabled market. Accordingly, in this chapter, we explain structure of the smart grid and discuss various authentication schemes associated with it. We have described different security parameters and varied attacks, which should be considered for successful and secure usage of the smart grid system.

Keywords Authentication · Data · Security · Smart Gird

4.1 Introduction

The development of the traditional electric grid can be dated back to the 1880s when the major source of energy were hydraulics, coal gas and steam. The first power station was built in United Kingdom by Charles Merz of the Merz & McLellan consulting partnership, named as Neptune Bank Power Station back in 1901 and had the largest integrated power system by 1912. As built in twentieth century, the traditional electric grid is basically an interconnected network, which supplies electricity from the production house to the consumers. The electric grid system architecture consists of basic three components:

1. Generating stations (power stations located near fuel sources) producing electrical power
2. High voltage transmission lines

T. Limbasiya (✉)
Birla Institute of Technology & Science (BITS), Pilani, Goa, India

A. Arya
NIIT University, Neemrana, Rajasthan, India

A. V. D. M. Kayem et al. (eds.), *Smart Micro-Grid Systems Security and Privacy*,
Advances in Information Security 71, https://doi.org/10.1007/978-3-319-91427-5_4

3. Distribution lines distributing power to individual customers

Power generation stations are located generally far away from the mainstream cities for easy production. These power stations send produced electricity through high-voltage transmission lines to the grid system. Grid systems get power from generation stations and provide power to different customers according to their requirements. This electricity is stepped up to a higher voltage before transmission for effective transmission. After that, it is associated to the electric power transmission network. Smart grid system has different advantages regarding major electricity supply schemes. We have a high impact of smart micro grid systems to make easy administration of various electricity demands with fewer human resources in technology enabled world. Because, smart grid technology can be installed at varied places by collaborating with many private or government organisations [4]. There is a vast progress in the development of technology day by day. By the increase as the technology, working patterns have been transformed from time to time and we should adopt new technology in the working system. If we do not take into consideration them, then we may fall into traps or find ourselves in a major problem or we cannot sustain current lifestyle effectively. Coming to the energy sector, it needs to use some novel idea and take advantages from it to reach the demand. A power/electric gird maintains need and supply of locality (residences, major manufacturers, colonies, etc.), which can be viewed as a micro grid. We can consider a micro grid as an electric grid for smaller areas whereas electrical grid can be taken for larger areas. Even though the architecture and the working of a traditional grid is well planned, it suffers from several drawbacks under different fields. Some of these drawbacks are discussed as follows:

1. **Age old equipment**
   The traditional electric grid systems dates back to the twentieth century and thus, there is a high failure rate, which affects performance at the consumer end. Therefore, it leads to high customer interruption rates. In order to ensure proper working of the equipments, it is essential to maintain them. As a result, the system requires higher maintenance cost.
2. **Outmoded system layout**
   The system layout in older areas require addition of substation sites, which is difficult to achieve and thus, they are enforced to rely on existing limited facilities.
3. **Obsolete engineering**
   Traditional power delivery planning tools and engineering systems are not worthwhile in today's scenario. Additionally, the outmoded layouts and aged equipments just aggravate problems.
4. **Old cultural values**
   The planning was beneficial in vertically integrated industry, which exacerbates problem under deregulated industry.

Different changes in daily lifestyle of individuals have been observed due to these drawbacks along with unprecedented advances in technology. Inability to adapt to this fast changing environment not only impacts professional lives of individuals but also affects them psychologically. A huge shift in mode of operations has been adopted different fields including the energy sector. In order to fulfill the demand in the energy sector, there is a need to deploy various novel approaches. This has led to the development of a smart grid technology, which is capable to solve encountered various problems with traditional grid systems and at the same time, it ensures secure data transmission.

4.1.1 Electric Grid to Smart Grid

Smart grid is a new paradigm evolved from an electrical grid. Electrical grid only deals with large transmission areas, but smart grids include all form of the electrical grid and incorporated with much automation in data information flow along with advanced energy delivery network. Based on the current survey, smart grid research is presently focused on infrastructure based systems, management systems, and protection-focused systems mainly [17]. For transforming from the traditional grid to the smart grid, we should replace physical infrastructure with digital infrastructure. An ideal smart grid needs to provide, observe, control, manage, protect components within the system. This needs to be made at high resolution in space and time. Technology is being upgraded from day to day, but the electric grid is still in the same phase. Electricity is the major and mandatory requirement that we want for our households, manufactures, and others. Modernising the traditional grid system is essential presently. This can be achieved by investing in new energy infrastructure called smart grid. It combines the power of traditional grids and automation technology with which we achieve information technology with power transmission for our benefit(s). There are numerous advantages of smart micro grid systems. Components of the smart grid system should ensure efficient working. However, the performance may vary according to available infrastructure and requirements of the customers.

1. **Intelligent Devices**
   These devices are deployed at the customer end which are capable of determining customers' requirements. This decision is taken based on some predefined data supplied by the customer. The deployment of these devices will not only help customers to reduce their power consumption but it also help in reducing peak load of overall power supply.
2. **Smart Meters**
   Power meters can be used to automate multiple functionality such as billing, detecting power failures, providing efficient, identification of tampering, faster customer care services, etc.
3. **Smart Distribution**
   This consists of self-healing, self-balancing and self-optimising, which are some of the key features associated with smart grids. These features facilitate long distance transmission, automated monitoring and analysis tools, which can be

used to detect and/or predict cable and power failures based on real-time data about environmental as well as past data.

4. **Smart Generation**
   In order to optimise the generation of electricity, it is essential to identify fluctuating demands and maintain the voltage, frequency and power factor standards. This relevant data is usually obtained from multiple points in the grid.

A smart grid system has different helpful characteristics, which are useful directly or indirectly to the electrical grid system. The characteristics are as the followings:

- Increase the usage of digital information and controls technology
- Dynamic optimisation
- Deployment and integration of advanced technologies
- Development and incorporation of multiple demands
- Deployment of smart systems
- Compilation of smart appliances and consumer devices
- Time effectiveness
- Identification and lowering of unreasonable or unnecessary barriers

4.1.2 Smart Grid to Smart Micro Grid

A smart micro grid is a modified version of an electric grid which does not only focus on producing bulk electricity but it also employs various operational and measures energy using smart meters, smart appliances, renewable and energy efficient resources. It not only focuses on a one way distribution of electric power like the traditional electric grid but this new smart grid practices two-way flow of electricity and information and this helps in optimum and advanced distribution of electrical energy. In order to understand the smart grid, it is essential to first make sense of the traditional electric grid structure. The traditional electric grid can be viewed as a mesh network constructed for distribution of power from the suppliers to the consumers. Several generation stations provide power to these grid systems, which is then transmitted to different customers depending upon their requirements.

Figure 4.1 shows a general architecture of smart micro grid systems, which contains different components for various purpose(s). These electricity generating power stations are generally located at remote locations from heavily populated areas. Since the electric power is to be transmitted to a large distance, it is first stepped up to a higher voltage and then carried out by high voltage transmission lines. These lines provide electricity to all customers, who are none other than the local electric power distributor. On arrival at the substation, the voltage is stepped down from a transmission to distribution level. Furthermore, the voltage is again converted from distribution level to essential service voltage.

A smart micro grid system includes numerous benefits. Smart grids combine strengths of the traditional grid system along with automation technology, which

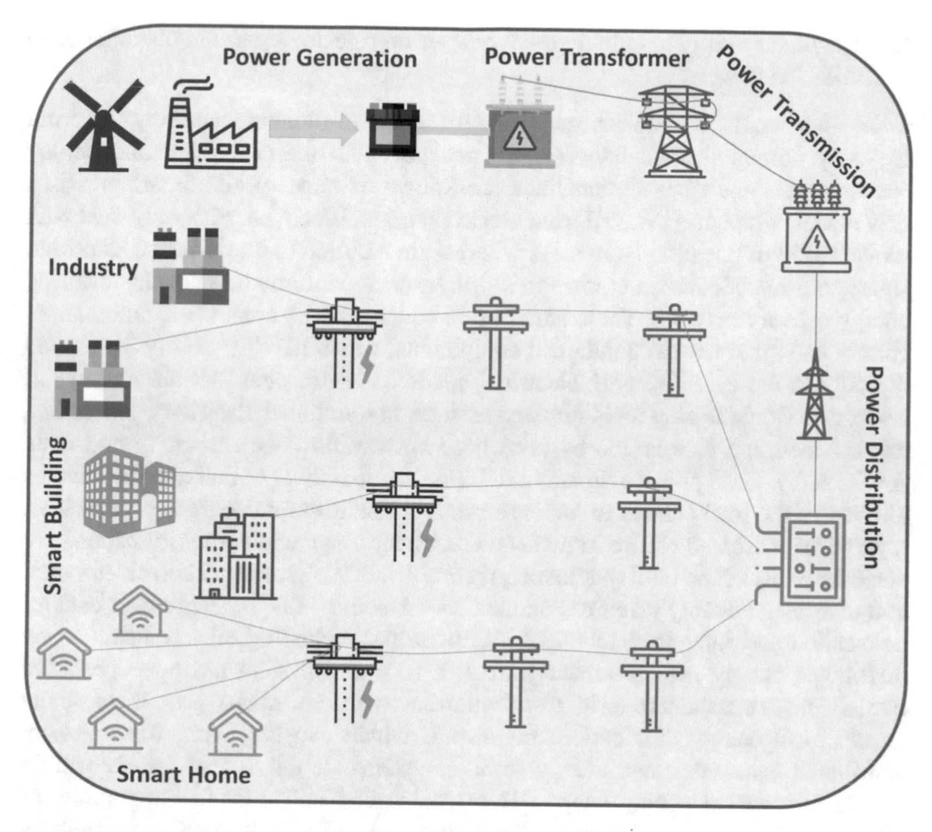

Fig. 4.1 A general architecture of micro smart grid system

helps us achieve information-centric power transmission for overall benefits. Some of the advantages associated with smart grids are discussed as follows:

1. They eliminate the usage of ageing equipments encountered in the traditional grid systems. The transformation from traditional to smart grid system involves replacement of physical infrastructure with digital infrastructure.
2. They use grid adequately to meet the increasing demand.
3. They decrease frequency of brownouts, blackouts, and surges.
4. They provide facility of sensing in the transmission lines.
5. Smart grids makes lower energy costs as they have ability to readjust the amount of transmission according to the customer's needs.
6. Customers get control over their power bill.
7. Real-time troubleshooting is facilitated.
8. Expenses borne by the energy producers are reduced.
9. Broad-scale electric vehicle charging is facilitated.
10. They have potentiality to be consolidated along with renewable energy sources on a large level, which leads to load sharing and reduction.

11. They have ability to quickly recover after unexpected breakage/disturbance in lines and feeders.

In other words, we can say that grids make use of digital technology to permit two way communication between the producer and the consumer, and deploy sensing along the transmission lines are known as smart grids. Smart grid is a new model, which has evolved from electrical grids. Electrical grids only deal with transmission of power in large areas whereas, in addition to transmission of power, smart grids include many automation's with respect to information flow in advanced energy delivery network. These smart grids equipped with computers, automation, control and other new technological equipments, which have the ability to respond digitally to the ever changing electrical needs. It is expected that the concept of smart grid would lead to revolutionary changes in traditional electricity generation, transmission and distribution by allowing two-way flow for electricity and data. Additionally, these power generating units are comparatively small and they can be allotted to the load centres to increase reliability and to reduce transmission loss. It provides increased choice to manage electricity usage and ability to respond for electricity cost adaptation by adjusting their utilisation. A smart grid involves varied and distributed energy resources and satisfies electric vehicle charging. It enables connection and integrated functions. With further technological advancement on distributed energy resource management, a new grid system has been proposed namely the smart micro grid distribution network. A micro grid is basically an electrical energy distribution framework, which consists loads, transmission, distributed generators, and energy storage systems. A micro grid can frequently update changes in energy supply. Here, controlled and reliable integrations of distributed energy resources and micro grids are exceedingly essential to guarantee a continuous power supply in the most systematic and economic framework [20, 22].

In this chapter, we survey various authentication models associated with smart micro grid systems, and identify potential security attacks for the same. Then, we explain different security attacks in detail to recognise the significance in the advanced micro grid technology.

Chapter Organisation: In Sect. 4.2, we present a brief survey regarding various existing smart micro grid authentication schemes. In Sect. 4.3, we describe security challenges and various attacks, which are associated with smart micro-grid authentication models. Finally, we conclude our chapter in Sect. 4.4. References are at the end.

4.2 Related Works

Generally, there are different security concerns such as availability, authenticity, confidentiality, integrity, and privacy. A lot of dominant remote user authentication systems provide genuineness based on different factors (text password, smart card, biometric identity). However, Lamport introduced the first authentication system to prevent a system from numerous challenges, which were focused on one-factor authentication [1]. Researchers recommended varied solutions to resist the smart grid system against different attacks. They discussed authentication methods based on diversified aspects through which systems can be protected to provide a certain level of security. At the same time, there are some drawbacks related to illegal access in authentication schemes, which led to the system vulnerable [24].

4.2.1 Traditional Electric Grid System

In 2008, Hamlyn et al. [2] worked on a utility computer network security management and authentication system, which was responsible for action and command requests in smart grid operation. The proposed system offered a starting point for securing and authenticating the smart grid networks. While focusing on the new security architecture design for smart power grids, the management worked upon multiple security domains. The paper presented strategies and procedures for security checks as well as authentication of command requests in the host area electric power system (AEPS) and interconnected multiple neighboring AEPS. In 2008, Xu et al. [3] presented a hash tree based authentication technique in which the time delay was approximately ten times lower than RSA. Due to this lower time delay, the scheme proposed by Xu et al. [3] proved to be more time efficient and useful for providing authentication in home application network.

In 2009, Bobba et al. [5] focused on building a policy and key encapsulation mechanism-data encapsulation mechanism (PKEM-DEM) encryption scheme which was resilient towards chosen cipher text attacks. Moreover, they developed a policy based encryption system (PBES) using the scheme that provided these capabilities. They implemented PBES and measured its performance. In 2009, Valli et al. [6] realized that the pre-existing electrical grids are a closed system. Thus, they explored the various security issues and problems that could potentially arise if these smart meters are added to the existing grid connections.

4.2.2 Smart Grid and Micro Smart Grid Systems

In 2010, scientists [9] realized that in order to update the traditional grid systems, improved grid distributed intelligence and broadband communication capabilities will be essential as they will play a significant role in ensuring efficient operation of the updated grid systems. Due to this, the need for latest security technology for large wide-area communication networks had arisen and it leads to a discussion on the key security technologies including public key infrastructure and trusted computing. As a result, several researchers started working on schemes to protect the smart grid networks. In 2010, authors [10] suggested a network protocol to protect the power grid automation against cyber attacks.

Bennett et al. [11] tried introducing malevolent agents in order to study their effects on the network. According to their study, even though there were multiple inherent security concerned matters in wireless network. The most germane attack against the existing networking protocols used by AMI (Advanced Metering Infrastructure) is the black hole attack. Their work focused on how to avoid this vulnerability by establishing a faithful connection between source and a sink.

In addition, in 2011 Li et al. [13], and Nicanfar et al. [14] together thought usage of multicast authentication for one time signatures. And this reduced the storage overhead significantly. This permits in the signature key size reduction. In the same year, Fouda et al. [15] used Diffie-Hellman key agreement protocol (DHKAP) as an authentication scheme for home appliance network. After the analysis, it can be concluded that their scheme makes use of less memory and also reduces the communication overhead.

Later in 2012, Nicanfer et al. [16] deployed multilayer consensus based password authentication protocol and successfully managed to decrease system security overhead. They did this by employing a one-way hash function and accompanying it with a primitive password between the appliance and the HAN controller. Further, four state individual agreed on password-based authenticated symmetric key establishment between the appliance and the upstream controllers during transmission of 12 packets. This significantly enhanced security process, data delivery time and defended the system from various attacks such as replay, off-line password guessing, man-in-the-middle, compromised impact, ephemeral key compromise impression, and unknown key-share. On the very same path, Kim et al. [18] managed to reduce the amount of computation in the smart meter by employing matrix based homomorphic hash function. To prove the correctness of their concept, they did a complete security and performance analysis of the protocol.

In 2013, Nicanfer et al. [19] focused on reducing delay caused by the security process and proposed an authentication scheme, which was based on multilayer consensus based password authentication key exchange elliptic curve cryptography (MCEPAK). They further developed an enhanced identity based cryptography model, which reduced management overhead significantly. In order to secure the system from several security challenges, in 2014 Li et al. [23] suggested an authentication scheme focused on merkle hash tree technique. After the analysis, one can see that the authentication scheme has ability to protect against various attacks such as replay, message injection, message analysis, and modification. In 2015, authors [24] realized that the basic security principles such as confidentiality, integrity and availability cannot be overlooked. As a result, they discussed the importance and need of mutual authentication among all the entities comprising the smart grid and took care of all the security principles with lower overhead, delay and cost.

4.3 Security Challenges in Smart Micro Grid

Due to technology expansion and advancement in the smart grid systems, cyber security is an essential requirement. Stationing a smart grid system without any strong security measures (authentication, availability, confidentiality, integrity, and privacy) may lead to cyber-attacks. Security measures have been implemented, but if they are not up to the level of current technology, then it may lead to compromising the entire system and its stability. We must make sure that the security mechanisms need to be enforced with proper authorisation to get access to the smart grid management system [20–22].

Ensuring that the operations of any system are secure and protected against various kinds of attacks and it is an integral part of system development. In order to achieve this, we need to make sure that the system is fulfilling three basic requirements of security—Confidentiality, Integrity (of both data as well as device) and Availability. In this section, we elaborately explain different requirements of security from the perspective of a smart grid network.

1. **Accountability**

   Users face various security challenges in the smart grid even though new emerging and advanced cyber security technologies are trying to protect the smart micro-grid infrastructure. Recently developed advanced systems monitor all incoming and outgoing electricity flows. It also keeps data about power consumption and generation in home areas. At the same point, a system should take care about data updating of various fields. Thus, the reading updated by the smart meter should ensure accountability that data is from the specified smart device only. Otherwise, a system might not response correctly regarding data incidents. Figure 4.2 displays an accountability requirement in the smart micro-grid framework. Accountability is one of the crucial properties, which says that the action against an entity can be traced back to the entity. It means if there are ten users in the system, then every user should be held accountable for his actions. Similarly three users (in Fig. 4.2) are accessing the server, and they might be changing, managing and updating data in the server. To ensure accountability, one should be able to trace that which user did which changes and should be held accountable for it.

2. **Authenticity**

   Since the smart grid network consists of millions electronic appliances and users, an authenticity is the process for confirming the identity of a device or a user as a precondition for ensuring the authorisation to the resources in the smart micro-grid. A hard-and-fast access control must be practiced to restrict unauthorised users from accessing sensitive data and controlling critical infrastructure(s).

   To ensue authenticity, the smart grid should include fundamental verification system to check a user and to ensure data Protection. Authenticity is one of the

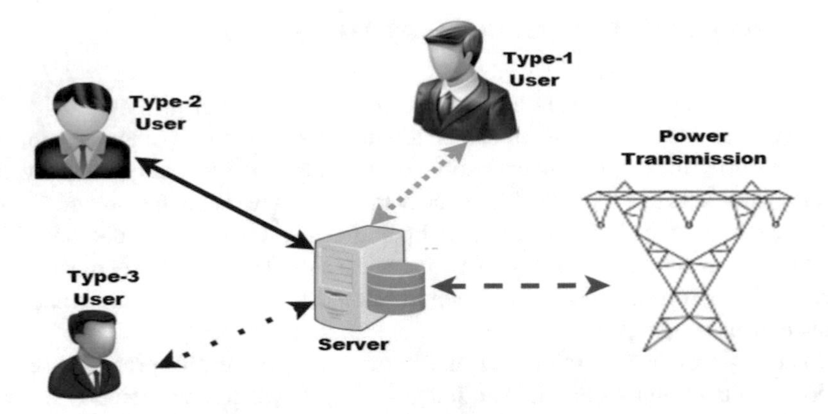

Fig. 4.2 Accountability in the smart micro-grid mechanism

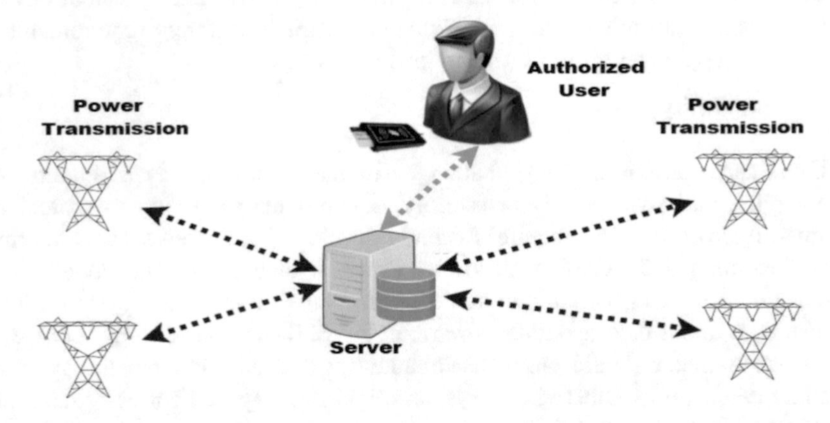

Fig. 4.3 Authenticity in the smart micro-grid model

most critical security principals and we should ensure it. As shown in Fig. 4.3, the smart micro grid communicates with many smart devices and users. At the same point, we must deploy an authentication scheme to ensure every single smart device, which is trying to connect in the smart micro grid.

3. **Availability**

Availability refers to on time resource(s) arrangement from the system when a legitimate user is trying to access it. This means that both (system and information) should be timely available to the user irrespective of time. In the context of smart grid system, we are concerned about availability of the smart grid meter and control system. These components are easily susceptible to a denial of service attack and hence, the legitimate user does not get services from the system. Since the smart grid systems are based on wireless communication technology, attackers try to add noise to signals, which leads to a jamming attack.

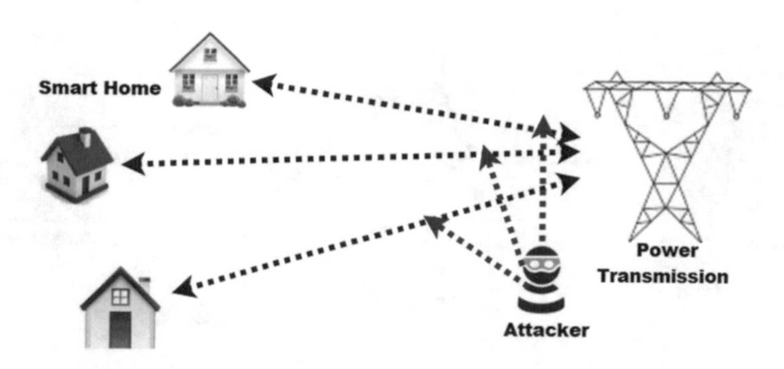

Fig. 4.4 Availability in the smart micro-grid framework

This is usually achieved by sending same frequency noise signal, which interferes genuine signals. Hence, services will be unavailable and the system cannot function according to requirements. It successfully deteriorates transmitted data signal, making it impossible for the user to get information from the distorted packets. The smart micro grid should not face a downtime and it should have maximum availability even though more number of attackers would attempt to prevent legitimate users from connecting to the smart grid communication infrastructure to obtain various services. As seen in Fig. 4.4, legitimate users are connected and an attacker is trying to bring down availability.

4. **Confidentiality**

Confidentiality ensures that all the property and personal information are accessed only by the legitimate and authorised person. Confidentiality can be expressed in terms of both components as well as data. If confidentiality is ensured in the system, then we can assure to users about the system that no malicious user can access to others' confidential data. In a smart grid system, confidential information such as power usage of a particular customer, customer's registered account information, etc. are linked with each customer's account. Many attackers with malicious intention try to eavesdrop on communication channels to get this information. Even though access to this information does not directly impact on the smart grid system working, it can lead to breach of users' privacy.

Researchers claimed that by accessing consumption usage via the smart meters, attackers can identify all smart electrical appliances that a particular customer makes use of it. Accordingly, attackers can easily draw inferences about financial stability of the user, their behaviours and preferences [7]. Several models [8] have been proposed in order to preserve customers' privacy in a smart grid. Nowadays, many peer-to-peer protocols have been implemented in which smart meters collect usage data for billing purposes, supporting load balancing and various monitoring functions to preserve consumer's preferences. Figure 4.5 shows confidentiality requirements in the system. Confidentiality

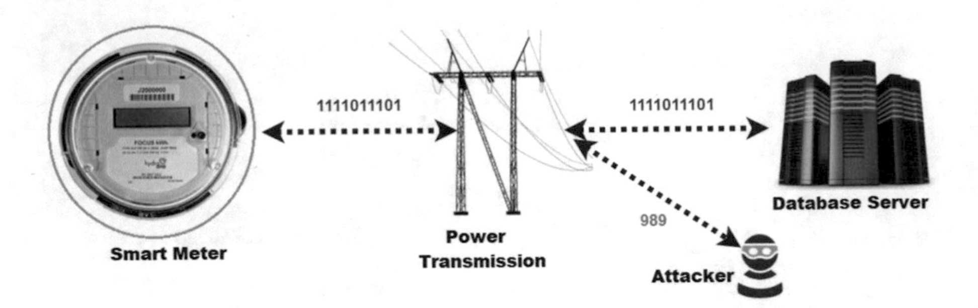

Fig. 4.5 Confidentiality in the smart micro-grid scheme

is very significant because data is being transferred from a smart meter to the database server. While transmitting this data, we should ensure that data is only understandable by a legitimate person/system. We can observe that an attacker is trying to read confidential information in Fig. 4.5 and s/he can disclose various kinds of information.

5. **Integrity**

Integrity refers to the fact that information available at the receiver end is not altered. In other words, it can be said that no attacker with malicious intentions has been succeeded in manipulating or modifying information. There are several pieces of critical information such as smart meter readings, control commands, billing information, etc. and this data is associated with the smart grid network. Integrity of the smart grid network can be easily compromised if any of this information has been manipulated by an attacker. In 2011 Liu et al. [12] discovered a class of attacks known as false data-injection attack, which can be easily introduced in the electrical power grids. If an attacker is successful in compromising few smart meters, he or she can easily access power system configuration information. As a result, fake or manipulated data can be easily injected into the monitoring centre.

There are various ways to ensure integrity, i.e., certification and authentication. It is essential to note that any substation and linked smart devices should be able to authenticate and validate each other's identity to prevent impersonation. By employing data certification, we can safeguard messages from modification during transmission. Substations employ attestation to confirm memory contents, which include data and code available on smart device have not been altered. Public key cryptography ensures security related to integrity but adoption of public key cryptography further requires a trusted third party for key management. Integrity is a very vital security principle, which ensures data accuracy. It means that transmitted data should not be modified by an attacker. If it is done, then integrity of the data is compromised. In Fig. 4.6, one can be seen that data is being transmitted from a smart meter to the server and it has been altered by an attacker. For an example, transmitted data was 1111011101, and an attacker has compromised this data to 1100101111.

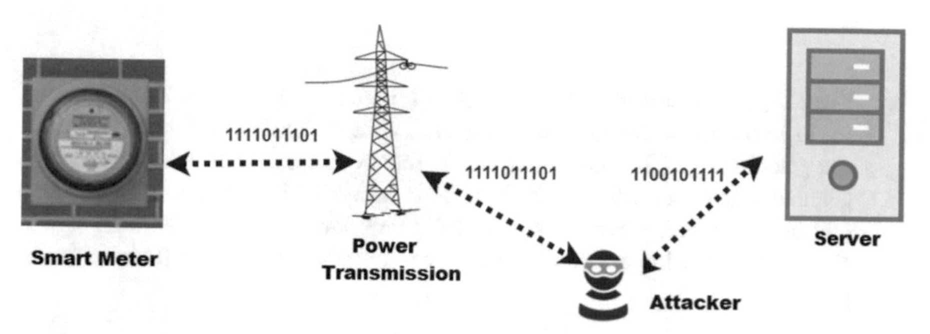

Fig. 4.6 Integrity in the smart micro-grid structure

4.3.1 Security Attacks in Smart Micro-Grid Systems

Before providing an authentication scheme for the smart grid, one should be aware of various possible attacks in the smart grid system. Attacks are described in details as follows.

1. **Denial of service attack**

   A denial of service attack occurs when legitimate users do not receive any services or resources after requesting for the same. This happens if the service is not available and its resources are occupied. As a result, a legitimate user wants to avail the service but he/she does not get due to unavailability. Considering this situation in the smart micro grid network, the smart meter, network devices and communication link servers can be severely affected by an attack. Further, the control of the electricity is affected and this leads to stopping the power supply. If a legitimate user's electricity power supply is tampered or cut, then it is not fair because s/he pays Particular amount for it.

2. **Man-in-the-middle attack**

   An attacker tries to intercept data, which is being transferred via a telecommunication medium. Then, the sender thinks that an attacker is the receiver and the receiver thinks that an attacker is the sender. This gives an opportunity to an attacker to forge data easily and manipulate messages. In the context of smart grid, a man-in-the-middle attack can be launched when the communication channel is jammed by inserting false data and further, it delays the transmission. Consider a scenario where confidential information (the units of electricity used by a particular user) is being sent via a public communication medium. If this kind of information is intercepted by an attacker, then he/she can easily estimate the standards of living of the user and probably plan something undesirable and malicious against him/her such as theft.

3. **Jamming attack**

   Since the smart grid is based on wireless communication technology, attackers try to introduce noise in signals and therefore, it leads to a jamming attack. They achieve this by sending same frequency noise signal, which interferes genuine signals and thus, it leads to lack of availability and its performance will be decreased. Hence, it successfully deteriorates transmitted data signal. Hence, it makes impossible for the user to correct information from distorted packets.

4. **Distributed denial of service of attack**

   This is similar as a denial of service attack but in this case attackers practice multiple systems to target one legitimate system. The probability of occurrence of this attack in the smart meter is high. Because the embedded device based AMI provides service via CPU can be easily exploited.

5. **Phishing attack**

   Phishing is an attack in which unauthorised users steal users' personal information by sending a fake request to them and pretending to be the original concerned authority. They provoke user to response their sensitive credentials. The smart grid has many home appliances in the home network system. A successive phishing attack could lead to the revelation of keys.

6. **Modification attack**

   Energy is the nation's property and a theft of it is a big crime. If an attacker has an opportunity to alter communicated data from one end to another point, then it is a big challenge to system in terms of security requirements. As a result, a service provider will lose electricity power. Sometimes, adversaries can tamper smart meters, which is done in order to steal electrical energy. Therefore, a smart meter shows less consumed electrical power and consumers should pay less amount.

7. **Spoofing attack**

   Spoofing is a kind of attack in which a person or program masks essential data to get the illegitimate privileges. This kind of activity can be possible in the remote user applications because both (client and server) are placed at different locations and they should communicate over a public communication medium. If an adversary (behalf of a legal user) can send a forged login request to the server and he/she succeeds to get the authorisation for various privileges, then it will be lose to the smart grid system because an attacker can perform different activities after getting the access from the system.

8. **Plaintext attack**

   Before understanding a plain text attack, one should understand what do you mean by a plain text. If data is transferred into a human readable form from one end to other points, then it is susceptible to a plain text attack as it makes easy for an attacker to read the data. When we talk about smart micro grid infrastructure, more number of smart devices are associated with each others in the infrastructure. A smart home has many smart appliances, which are connected to the smart meter and it keeps record and sends it to the smart micro grid server. While this confidential data is being transmitted, we should ensure that it is not transferred in normal form but, a cryptographic technique should be applied to this data before disseminating it. If a sender sends data without performing an encryption process, then data is susceptible to a plain text attack.

9. **Insider attack**

   In general, users make usage of the smart micro grid system at common area. Therefore, a neighbour might have little knowledge about how the infrastructure works and a little knowledge about how the authentication works and what user's credentials might be? An adversary can be internal person, who belongs to the system. Additionally, s/he can obtain different data by intercepting packets (transmitted through a public medium) while the system verifies generally and thus, he/she can successfully launch an insider attack.

10. **Leak of verifier attack**

    An authentication mechanism generally involves three parties, which are a user (which needs to be authenticated), the object (like a server or database for which the user needs to be authenticated) and a verifier (which authenticates the user). Suppose the verifier consists of all details of a user and his/her verification key(s) and the verifier gets compromised by any other authority or a person, then the entire system can be settled down according to an adversary's requirement(s). Similarly, a smart micro grid user tries to be checked by the verifier, at the same point, if the verifier server is compromised, then confidential data (user name, password, some computed parameters, etc.) would be accessible to an attacker easily, which leads to settlement of the whole smart micro grid infrastructure.

11. **Impersonation attack**

    An impersonation attack can be defined as when an unauthorised user attempts to mimic as a legitimate user and attempts to access various privileges illegally. A million of smart homes are using smart micro grid and each of these users have an account as well as each of these accounts get authenticated by the server. If an attacker intercepts these packets over an open channel, then an attacker might get to know user's credentials. Furthermore, he/she does analysis and if obtains correct credentials, then s/he can apply an impersonation attack without knowledge of an original person.

12. **Temporary information attack**

    This kind of attack is applied to in certain circumstances, in which an attacker targets the highly configured system or the high-ranking authority of the smart micro grid framework. Therefore, the mechanism designs a specified level of security for essential authority/system. Additionally, the entire smart micro grid structure is controlled by the concerned dominance. If an attacker obtains the secret key or related credentials and s/he can intercept the communication system, then it will be beneficial for an attacker to break down the system. Further, an adversary can damage relative architecture, which leads major problem in the smart micro grid system.

13. **Session key disclosure attack**

    If any legal user wants to access the smart micro grid system, then a user should send a login request with valid credentials, and the server checks a message request. If it is reasonable, then the server and a user generate a session key based on mutual parameters, which is valid for a limited period only. However, an attacker can get various credentials from a public communication channel and attempts to calculate a session key. If an adversary succeeds in this process, then s/he has an opportunity to apply a session key disclosure attack.

4.4 Conclusion

We have illustrated the smart grid structure and its advantages in the real-life scenario. After that, we have discussed different user verification systems in the smart grid framework, which can be implemented in the concept of smart micro grid mechanism. Next, we have described fundamental security needs for the smart grid system and the structure can be damaged directly and/or indirectly if these requirements have not been achieved. Furthermore, we have familiarised various security attacks, which can be applied in the system by an adversary through varied methodologies. At the same point, a user or the system may lose different significant credentials, which may affect immediately or later to the user or system directly or incidentally.

Acknowledgements We are grateful to the anonymous reviewers for their noteworthy time and effort in suggesting their exceptional views on the chapter in the review process. In addition, we are thankful to editors for the handling of this book. This work is partially supported by NIIT University, Neemrana, India.

References

1. Lamport, L. (1981). Password authentication with insecure communication. *Communications of the ACM*, 24(11), pp. 770–772.
2. Hamlyn, A., Cheung, H., Mander, T., Wang, L., Yang, C., & Cheung, R. (2008). Computer network security management and authentication of smart grids operations. *In Power and Energy Society General Meeting-Conversion and Delivery of Electrical Energy in the 21st Century*, IEEE, pp. 1–7.
3. Xu, K., Ma, X., & Liu, C. (2008). A hash tree based authentication scheme in SIP applications. In Communications, *ICC'08. IEEE International Conference on*, pp. 1510–1514.
4. Kaplan, S. M. (2009). Electric power transmission: background and policy issues. *US Congressional Research Service*, April, 14, pp. 4–5.
5. Bobba, R., Khurana, H., AlTurki, M., & Ashraf, F. (2009). PBES: a policy based encryption system with application to data sharing in the power grid. *In Proceedings of the 4th ACM international symposium on information, computer, and communications security*, pp. 262–275.
6. Valli, C. (2009). The not so smart, smart grid: Potential security risks associated with the deployment of smart grid technologies. *In Proceedings of the 7th Australian Digital Forensics Conference*, pp. 19–23.
7. Quinn, E. L. Privacy and the new energy infrastructure, (2009).
8. Cavoukian, A., Polonetsky, J., & Wolf, C. (2010). Smart privacy for the smart grid: embedding privacy into the design of electricity conservation. *Identity in the Information Society*, 3(2), pp. 275–294.
9. Metke, A. R., & Ekl, R. L.(2010). Security technology for smart grid networks. *IEEE Transactions on Smart Grid*, 1(1), pp. 99–107.
10. Wei, D., Lu, Y., Jafari, M., Skare, P., & Rohde, K. (2010). An integrated security system of protecting smart grid against cyber attacks. *In Proceedings of IEEE Innovative Smart Grid Technologies*, pp. 1–7.
11. Bennett, C., & Wicker, S. B. (2010). Decreased time delay and security enhancement recommendations for AMI smart meter networks. *In Proceedings of IEEE Innovative Smart Grid Technologies*, pp. 1–6.
12. Liu, Y., Ning, P., & Reiter, M. K. (2011). False data injection attacks against state estimation in electric power grids. *ACM Transactions on Information and System Security (TISSEC)*, 14(1), 13.
13. Li, Q., & Cao, G. (2011). Multi-cast authentication in the smart grid with one-time signature. *IEEE Transactions Smart Grid*, 2(4), pp. 686–696.
14. Nicanfar, H., Jokar, P., & Leung, V. C. (2011). Smart grid authentication and key management for unicast and multicast communications. *In Proceedings of IEEE Innovative Smart Grid Technologies Asia*, pp. 1–8.
15. Fouda, M. M., Fadlullah, Z. M., Kato, N., Lu, R., & Shen, X. S. (2011). A lightweight message authentication scheme for smart grid communications. *IEEE Transactions on Smart Grid*, 2(4), pp. 675–685.
16. Nicanfar, H., & Leung, V. C. (2012). Smart grid multilayer consensus password-authenticated key exchange protocol. *In Communications (ICC), 2012 IEEE International Conference on*, pp. 6716–6720.

17. Fang, X, Mistra, S., Xue, G., & Yang, D. (2012). Smart Grid - The new and improved power grid: A survey. *IEEE communications surveys & tutorials*, 14(4), pp. 944–980.
18. Kim, Y. S., & Heo, J. (2012). Device authentication protocol for smart grid systems using homomorphic hash. *Journal of Communications and Networks*, 14(6), pp. 606–613.
19. Nicanfar, H., & Leung, V. C. (2013). Multilayer consensus ECC-based password authenticated key-exchange (MCEPAK) protocol for smart grid system. *IEEE Transactions on Smart Grid*, 4(1), pp. 253–264.
20. Wang, W., & Lu, Z. (2013). Cyber security in the Smart Grid: Survey and challenges. *Computer Networks*, 57(5), pp. 1344–1371.
21. Zeadally, S., Pathan, A. S. K., Alcaraz, C., & Badra, M. (2013). Towards privacy protection in smart grid. *Wireless personal communications*, 73(1), pp. 23–50.
22. Bari, A., Jiang, J., Saad, W., & Jaekel, A. (2014). Challenges in the smart grid applications: an overview. *International Journal of Distributed Sensor Networks*, 10(2), pp. 1–11.
23. Li, H., Lu, R., Zhou, L., Yang, B., & Shen, X. (2014). An efficient merkle-tree-based authentication scheme for smart grid. *IEEE Systems Journal*, 8(2), pp. 655–663.
24. Saxena, N., & Choi, B. J. (2015). State of the art authentication, access control, and secure integration in smart grid. *Energies*, 8(10), pp. 11883–11915.

Chapter 5
A Resilient Smart Micro-Grid Architecture for Resource Constrained Environments

Anne V. D. M. Kayem, Christoph Meinel, and Stephen D. Wolthusen

Abstract Resource constrained smart micro-grid architectures describe a class of smart micro-grid architectures that handle communications operations over a lossy network and depend on a distributed collection of power generation and storage units. Disadvantaged communities with no or intermittent access to national power networks can benefit from such a micro-grid model by using low cost communication devices to coordinate the power generation, consumption, and storage. Furthermore, this solution is both cost-effective and environmentally-friendly. One model for such micro-grids, is for users to agree to coordinate a power sharing scheme in which individual generator owners sell excess unused power to users wanting access to power. Since the micro-grid relies on distributed renewable energy generation sources which are variable and only partly predictable, coordinating micro-grid operations with distributed algorithms is necessity for grid stability. Grid stability is crucial in retaining user trust in the dependability of the micro-grid, and user participation in the power sharing scheme, because user withdrawals can cause the grid to breakdown which is undesirable. In this chapter, we present a distributed architecture for fair power distribution and billing on micro-grids. The architecture is designed to operate efficiently over a lossy communication network, which is an advantage for disadvantaged communities. We build on the architecture to discuss grid coordination notably how tasks such as metering, power resource allocation, forecasting, and scheduling can be handled. All four tasks are managed by a feedback control loop that monitors the performance and behaviour

A. V. D. M. Kayem (✉) · C. Meinel
Hasso-Plattner-Institute, Faculty of Digital Engineering, University of Potsdam, Potsdam, Germany
e-mail: anne@mykayem.org; christoph.meinel@hpi.uni-potsdam.de

Stephen D. Wolthusen
Department of Mathematics and Information Security, Royal Holloway, University of London, Egham, Surrey, UK

Norwegian Information Security Laboratory, Gjovik University College, Norwegian University of Science and Technology, Trondheim, Norway
e-mail: stephen.wolthusen@rhul.ac.uk

A. V. D. M. Kayem et al. (eds.), *Smart Micro-Grid Systems Security and Privacy*, Advances in Information Security 71, https://doi.org/10.1007/978-3-319-91427-5_5

of the micro-grid, and based on historical data makes decisions to ensure the smooth operation of the grid. Finally, since lossy networks are undependable, differentiating system failures from adversarial manipulations is an important consideration for grid stability. We therefore provide a characterisation of potential adversarial models and discuss possible mitigation measures.

Keywords Resource constrained smart micro-grids · Architectures · Disadvantaged communities · Energy · Grid stability · Forecasting · Feedback control loop

5.1 Introduction

Energy use projections indicate that developing world consumption will rise by about 0.5% a year between 2010 and 2040 and by 2040, 65% of world energy consumption will be in the developing world [9, 23]. However, national power network capacity limitations make providing reliable access to power a challenging problem. Rural and remote areas are the most seriously affected by load shedding aimed at supporting more urbanised (industrialised) areas during power shortages. As such, these areas can remain isolated from the national grid, for long periods or when this is not economically viable, remain completely disconnected from the grid. Finding ways of addressing this issue can be helpful in economic development as well as enforcing the UN charter of human rights which lists "access to power" as one of the sustainable development goals. It is worth noting for instance, the electrification of households and small businesses in South Africa rose from 36% in 1994 to 70% in 2002, leading to a substantial boost to the economy [16, 36, 40]. By contrast, in rural areas and townships, energy theft remains a severe problem with up to 40% of generated power lost to theft [7, 9, 16, 38].

5.1.1 Context and Motivation

In this chapter, we focus on a computational model for enabling fair and equitable access to power in resource constrained regions.[1] We consider how micro-grids based on distributed renewable energy resource and storage systems can be organised to address the power provision problem in rural/remote areas while at the same time reducing the carbon blueprint on the environment. To this end, we propose supporting the micro-grid with a cyber system (smart micro-grid), to enable real-time monitoring and control of power generation, consumption, and transmission. This allows the micro-grid to offer reliable as well as efficient electricity man-

[1]This broadly describes rural and remote regions where regular access to power is hindered by technological limitations.

agement, dynamic pricing, and demand-response strategies [16]. For economic reasons, the smart micro-grid is conceptualised to handle grid management with a communication network formed of low cost computational devices. In order to distinguish this architecture from the standard micro-grid architecture, we will henceforth refer to it as a *Resource Constrained Smart-Micro-grid* (RSMG).

The RSMG can be described as an amorphous distributed model of users who agree to cooperate to share electricity by coordinating grid activities via a lossy communication network. Users (producers) with power generation units contribute excess unused power to the grid, and the grid coordinator must assign the power resources fairly to the users demanding power (consumers) and at the same time fetch the best price for the producers. This structure, coupled with the absence of a centralised grid monitoring and management facility requires a shift in conceptualisation from the standard grid model [16, 20, 32].

5.1.2 Problem Statement

Differentiating between failures due to network faults and ones dues to adversarial manipulation is a challenging problem. Straight-forward approaches to adversarially manipulating micro-grids include mis-recording power consumption and/or generation information and exposures of personally identifiable information as well as sensitive information. Adversarial manipulations result in failures on the part of the micro-grid to satisfy expected service level agreements and in a loss of user trust in the reliability as well the dependability of the system. Grid stability is tightly intertwined with reliability and dependability, therefore user withdrawals from grid participation due to a loss of trust in the grid's capabilities, can result in a breakdown of the grid. A further consideration is that the grid operates over a lossy communication network which is by nature unreliable and undependable. Correctly distinguishing between adversarial and normal failure situations is important not only in identifying cases of energy theft but also not to confuse network faults with deliberate attempts to subvert the operation of the network. These attacks may *inter alia* seek to destabilise the grid, cause generation and feed-in to be mis-recorded, or to reveal personally identifiable and otherwise sensitive information.

5.1.3 Contributions

We extend the reference model for RSMGs that Kayem et al. [17] to incorporate control network structures, and characterising possible adversarial behaviour as well as capabilities. Normal state estimation behaviour, is used as a reference monitor to distinguish normal from adversarial or faulty behaviours. In considering adversarial behaviour, we focus on the case of energy theft where the incentive for the user is to avoid paying for power consumption.

Due to the absence of trusted centralised grid monitoring and management system, we model our control network algorithms to run over a distributed system that may be subject to external attacks and particularly subversion. We consider four main aspects namely, *metering*, *power resource allocation (Demand Management)*, *forecasting* and *scheduling*. In *metering*, we consider two power consumption and/or generation data collection models. The first is a manual collection model, in which basically the household grouping device controller (user) physically inserts information about the household's consumption and/or generation. This approach is useful, in cases where the existing network devices and structures are not sensor configured to enable automated metering. In the second case, an automated approach is used where a snapshot algorithm is invoked periodically to capture the consumption and/or generation data. We note that it is possible to have a hybrid model that combines aspects from both the manual and automated models but for simplicity we will keep the discussion on how the manual and automated models of power data collection work. Power resource allocation (demand management) is aimed at allocating power in ways that adhere to a devised power distribution schedule, with the goal of ensuring grid stability, reliability, and trust. Distribution schedules are determined by forecasted power consumption patterns and on observations of grid performance. We model the operation of the smart micro-grid, from the control perspective, by using a feedback control loop to coordinate interactions between the various components of the control network.

5.1.4 Outline

We structure the chapter as follows. The Introduction (Sect. 5.1) provides the context and motivation for the proposed micro-grid architecture, we highlight why and how current smart micro-grid architectures must be extended to cope with RMGs. In Sect. 5.2 we discuss general existing work on power networks in general, moving on to smart grid architectures and finally to micro-grid architectures. We proceed, in Sect. 5.3 to describe the operation of our proposed reference model for distributed micro-grid architectures, focusing specifically on the structure of the network and on how the logical and physical layers communicate to ensure grid stability [2, 4]. In Sect. 5.4 we propose algorithms for the operation of the micro-grid architecture from the logical or control network point focusing on demand management and forecasting [39]. Section 5.5 characterises attack models focusing on the adversarial capabilities and behaviours used to facilitate energy theft and grid destabilisation from the control network point [3, 25–27]. We present our micro-grid architecture simulator in Sect. 5.6 and demonstrate with the help of our experimental results, the resilience of our proposed grid architecture to power theft attacks. Conclusions and suggestions for future work are offered in Sect. 5.6. The outline of the chapter is summarised as follows:

5.2 State-of-the-Art on Power Network Security

Reliable access to energy is a key enabler for any modern society as electric power networks not only underpin most critical infrastructures and services in some form, but also because even the risk of disruptions leads to costly, inefficient, and environmentally harmful mitigation measures. In the developing world, rural areas may either not be connected to national power networks at all, or be subjected to load-shedding as grid operators prioritise urban centres in case of generator capacity shortages [12, 16, 19, 38]. As a consequence, inefficient petrol or diesel generators are often used to satisfy electrical power needs [29, 31]. Distributed generation based on renewable energy sources can facilitate access to equitable access to energy whilst minimising negative and unsustainable impact resulting from this enhanced access that would arise if a conventional large-scale fossil fuel generation and long-distance transmission grids were followed [16].

5.2.1 Smart Grid Architectures

Distributed generation based on renewable energy sources can facilitate access to equitable access to energy in regions that are severely affected by load shedding or disconnected from the national power grid, whilst minimising negative and unsustainable impact resulting from this enhanced access that would arise if a conventional large-scale fossil fuel generation and long-distance transmission grids were followed [12, 16, 16, 19, 29, 31, 38]. A key concern when using renewable energy sources such as solar power, wind, biomass, and hydroelectric power is not necessarily their capacity, but rather uneven availability and capacity at different time-scales [21]. For wind turbines and photovoltaic generators, this can vary on a second-by-second basis. The standard procedure is to balance this by fast-acting conventional generators, which requires adequate generator capacity constituting a large fraction of the renewable generator capacity, or by compensating variations in renewable generation over larger geographic areas both of which require reliable transmission network capacity. However, the cost of providing such conventional generator and grid capacity particularly to rural and disadvantaged areas is likely to be prohibitive in addition to undesirable from an environmental perspective [11]. It is therefore important to ensure that demand and generation are closely matched in isolated micro-grids where the transmission capacity for load-following is insufficient [7, 33]. Whilst demand shaping has long been common for larger (industrial) loads, a volatile power network requires a more fine-grained level of control, but cannot necessarily rely on the use of smart meters and home automation networks (HAN) as are being deployed in higher-income areas. Therefore, an architecture that enables substitution of these components with low powered and/or re-purposed existing information technology components to offer power network state estimation and authenticated access to power networks, is essential. In this way, it would be possible to ensure reliable power for high-priority loads such as lighting or information and communication devices, while shifting or reducing lower-priority loads e.g. thermal loads such as boilers/geysers, to low consumption or demand periods [29, 31]. Scheduling and demand management can contribute to grid stability, lower costs to consumers, reduced reliance on fossil fuels, and reduced capital expenditure for grid and generator capacity. Moreover, despite significant efforts, access to national power grids is not cost-effective for remote, rural areas which stand to benefit most from the improved robustness, efficiency, and capacity of a micro-grid [21, 33]. Linking individual generators for back-up in case of load-shedding or failure or for non-electrified dwellings for improved reliability and availability of energy is a good strategy, but energy theft must also be addressed by controlling generation and load accordingly [1, 34, 35]. However, the standard centralised micro-grid management approach needs to be expanded to enable distributed and hierarchical state estimation [37]. This is important when micro-grids are designed to rely on distributed energy generation sources [5, 6, 11–14, 19, 28].

5.2.2 Smart Micro-Grid Architectures

A key concern when using renewable energy sources such as solar power, wind, biomass, and hydroelectric power is not necessarily their capacity, but rather uneven availability and capacity at different time-scales [21]. For wind turbines and photovoltaic generators, this can vary on a second-by-second basis. The standard procedure is to balance this by fast-acting conventional generators, which requires adequate generator capacity constituting a large fraction of the renewable generator capacity, or by compensating variations in renewable generation over larger geographic areas both of which require reliable transmission network capacity. However, the cost of providing such conventional generator and grid capacity particularly to rural and disadvantaged areas is likely to be prohibitive in addition to undesirable from an environmental perspective.

The other aspect to consider is power theft which often leads to overloaded power networks. Power theft not only destabilises the grid, but also discourages maintenance and user participation [11]. It is therefore important to ensure that demand and generation are closely matched in isolated micro-grids where the transmission capacity for load-following is insufficient [7, 33]. Whilst demand shaping has long been common for larger (industrial) loads, a volatile power network requires a more fine-grained level of control, but cannot necessarily rely on the use of smart meters and home automation networks (HAN) as are being deployed in higher-income areas. Therefore, an architecture that enables substitution of these components with low powered and/or re-purposed existing information technology components to offer power network state estimation and authenticated access to power networks, is essential. In this way, it would be possible to ensure reliable power for high-priority loads such as lighting or information and communication devices, while shifting or reducing lower-priority loads e.g. thermal loads such as boilers/geysers, to low consumption or demand periods.

Beyond this, energy needs are met by a variety of fossil fuel and biomass sources that are also not used in the most efficient manner, particularly in disadvantaged and rural areas. Particularly during colder months, water heating constitutes close to 30% of electricity demand, with other significant contributions from heating, indicating a significant potential for reducing peak demand by peak shifting, and increasing reliance on renewable energy—whilst aggregate consumption is relatively low in low-income, and especially informal dwellings, peak energy use is problematic and only imperfectly controlled by time-of-use tariffs [29, 31]. Scheduling and demand management hence can contribute to grid stability, lower costs to consumers, reduced reliance on fossil fuels, and reduced capital expenditure for grid and generator capacity. Moreover, despite significant efforts, access to national power grids is not cost-effective for remote, rural areas which stand to benefit most from the improved robustness, efficiency, and capacity of a micro-grid [21, 33].

Linking individual generators for back-up in case of load-shedding or failure or for non-electrified dwellings for improved reliability and availability of energy is a good strategy, but energy theft must also be addressed by controlling generation and load accordingly [1, 34, 35]. However, the conventional centralised approach needs to be expanded to enable distributed and hierarchical state estimation [37]. This is important when micro-grids are designed to rely on distributed energy generation source and where distributed control of the smart grid environments is necessary [5, 6, 11–14, 16, 19, 28].

5.3 Resource Constrained Smart Micro-Grid: Network Model

Our micro-grid model is built on a network of distributed generators and storage units situated at arbitrary points on the network. Grid stability is maintained by ensuring that all generated power is consumed, and maintaining an equilibrium between power generation and storage is reliant on balancing energy demand and consumption. We employ a producer-consumer (prosumer) model for matching electricity supply to demand to encourage consumers to shift consumption and/or generation to periods when this is beneficial to the stability of the micro-grid [39].

In line with our distributed architecture, the micro-grid control point is decided via a consensus algorithm [18]. The micro-grid architecture is supported by three network models namely, a *power network*, a *communication network*, and a *control network*. The power network handles the interconnections between power generating and consuming devices on the network, and ensures physical power use and storage. Power generation and consumption data is transmitted over a wireless network which we term the *communication network*. The *communication network* transmits all the data on power availability (generated), and consumption. Finally, the control network coordinates interactions between the power and communication network from the algorithmic perspective. The control network evaluates the data collected to ensure that the power network's objectives of stability and user trust are enforced. Controlling flows is helpful in for example increasing network stability and assuring maximum utilization of generated power. We discuss all three network models to provide a general overview of the micro-grid's operation.

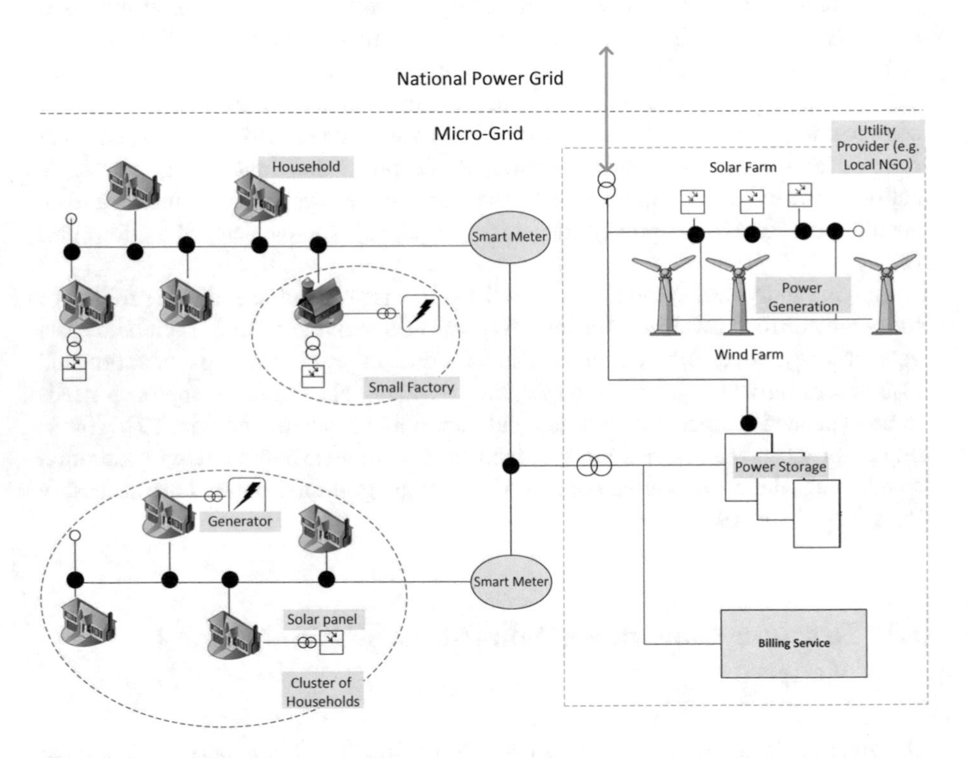

Fig. 5.1 Power network structure [17]

5.3.1 Notation

Let $\mathscr{A}$ be the set of household appliances such that a_i represents the ith appliance where $1 \leq i \leq |\mathscr{A}|$ is such that $|\mathscr{A}| \geq 1$. Each household contains at least one data aggregation unit $\mathscr{M}$ that handles power consumption and/or generation data collection at periodic intervals $\Delta t \in \mathscr{T} = 1, 2, \ldots, T$ where T is finite. The micro-grid consists of a set of C clusters of residential households and small businesses denoted by $C = \{c_1, \ldots, c_N\}$ where c_i represents the ith cluster and N the maximum number of household clusters on the micro-grid. Each household belongs in a cluster $c_j \in C$ such that $1 \leq j \leq |C|$ and C is the set of household clusters with a cardinality $|C| \leq N$ where N is the current maximum number of household clusters that the grid can support/sustain reliably. A smart meter SM is associated with c_j and each SM has a maximum nodal degree of d. The maximal nodal degree indicates the maximum number of households for which the SM can handle reporting accurately. A household is denoted by $h_{j,k}$ where j represents the SM to which the household is connected such that SM_j implies that $h_{j,k}$ is connected to the jth, SM where $1 \leq j \leq N$ with $N \geq 1$ indicating the current number of SM on the microgrid, and $k \geq 1$ but such that $1 \leq k \leq d$. So, for example, $h_{1,2}$ implies that the household is associated with SM_1 and is the household at position 2 in the cluster of households linked to SM_1. Each $\mathscr{M}$ is controlled by a set of authorised users $\mathscr{U}$ composed of subsets u_g, u_s, and u_r where u_g denotes users reporting power generation data, u_s users with power storage units, and u_r users providing power consumption reports.

5.3.2 Power Network

Our power network is designed to include households which are typically inhabited by single family units. Each household contains a set of electrical appliances that are each equipped with a sensor and connected directly to the power source. When an electrical appliance is not sensor-enabled, consumption must be reported manually. In the household, consumption and generation reports, are aggregated on a mobile device. A household may contain several mobile devices, but only one may serve as the household power aggregation device. Additionally, as shown in Fig. 5.1, each household cluster c_i linked to a shared smart meter SM_i. Sharing smart meters is a cost-effective solution for deploying smart micro-grids in economically challenged areas. Household power consumption and/or generation reports are transmitted from the mobile devices to the smart meter where they are aggregated for billing purposes.

Distributed Energy Resources (DER) handle the generation, storage, and physical delivery of electrical power to consumers. The power network can operate in either an island or grid-connected mode, but for simplicity we assume that the micro-grid operates in island mode. As illustrated in Fig. 5.1, we assume that grid participants can have various electricity generation units that are used to produce power to support the grid. Generated power is made available, at a cost, to grid participants and the excess generated power is stored in anticipation of future demand.

The power network operates on a producer-consumer (prosumer) model, in which a household can be both a power producer (when a generator is attached to the household) and/or a power consumer. There are two possible power network configurations that we envisage. In the first case, we assume that there exists at least one utility provider with the generation capacity to power the entire micro-grid, or least with a significantly higher proportion of power than the other users in the micro-grid. This is the scenario depicted in Fig. 5.1. All other users wanting to generate and sell electricity on the grid must sell to the Utility Provider who will then market the power to the users requiring power on the grid. The procedure for handling power marketing on the micro-grid is beyond the scope of this paper.

The second case, and this is the one we focus on in this paper, assumes a fully distributed model where every user can at the same time be a producer and consumer. In this case, the metering unit controlling the cluster of households, will periodically advertise, based on network observations power availability and will distribute the power to the houses in the cluster requiring this. Excess power will be advertised to other clusters on the grid. For simplicity we will use the term *Coordinator* from henceforth to refer to the *Grid Coordinator* or Utility Provider.

Each household $h_{j,k}$ contains a set of electrical appliances $\mathscr{A}$ such that $|\mathscr{A}| \geq 1$ and $a_i \in \mathscr{A}$ denotes the ith appliance in the household. In addition, $S(a_i)$ is the sensor associated with a_i and from which the consumption or generation report for a given period $\mathscr{T}$ is transmitted to the aggregation (mobile) device $\mathscr{M}$. Power consumption reports are stored on $\mathscr{M}$ using a vector representation, such that $S(a_i) \in \mathscr{M}\left(h_{j,k}\right)$ represents the value read from a_i during Δt for $h_{j,k}$. We express the vector representation as follows:

$$\mathscr{M}\left(h_{j,k}\right): \ [S(a_1)], \ldots, [S(a_i)], \ldots, \left[S\left(a_{|\mathscr{A}|}\right)\right] \tag{5.1}$$

In this data structure each slot contains the value transmitted from $S(a_i)$ where $1 \leq i \leq |\mathscr{A}|$ and $|\mathscr{A}|$ is the cardinality of the household appliances. We note that $|\mathscr{A}|$ is bounded by $|\mathscr{MAX}|$ the maximum number of appliances that households linked to $SM(j)$ can have.

For security reasons, a household can only declare a single $\mathscr{M}$ and can only report consumption or generation data from the household to which it is associated. Each household declares a set of users $\mathscr{U}$ who are authorized to control $\mathscr{M}$ and only one user at any given reporting period Δt can make reports. All changes must be declared explicitly. We further sub-divide $\mathscr{U}$ into three subsets namely, the subset of users with generation units, denoted u_g, u_s users with energy storage facilities, and u_r the users with neither storage nor generation capacity. We note that a user can belong both to u_g and u_s, but users in u_r cannot be in either one of or both u_g and u_s. If $u_i \in \left\{u_g \wedge u_s\right\} \vee \left\{u_g \vee u_s\right\}$ then $u_i \notin u_r$ and if $u_i \in u_r$ then $u_i \notin \left\{u_g \vee u_s\right\}$. Furthermore, a household belongs in a cluster of households, such that a household cluster c_j represents households with a common smart meter. Each metering unit SM has a maximum connectivity degree d such that $c_j = h_{j,1}, \ldots, h_{j,i}, \ldots, h_{j,|d|}$ and $h_{j,k}$ denotes the kth household in the jth cluster of the micro-grid. As shown in Fig. 5.2, the data structure at SM is a two-dimensional $|d| \times |\mathscr{MAX}|$ matrix where, each row in the matrix represents the power system state for $h_{j,k}$ during Δt. The smart meters agree on the coordinating smart meter for grid control operations. In the initial phase, the coordinating smart meter is randomly selected and the duration period T_{op} agreed on. On termination of the period T_{op}, a leader election algorithm [18] is initiated to decide on the next coordinating smart meter. The discussion of how the leader election algorithm operates is outside the scope of this paper, but

$$\mathcal{SM}(c_j): \begin{vmatrix} [h_{j,1}, S(a_1)], & ..., & [h_{j,1}, S(a_i)], & ..., & [h_{j,1}, S(a_{|\mathcal{A}|})] \\ ..., & ..., & ..., & ..., & ... \\ [h_{j,k}, S(a_1)], & ..., & [h_{j,k}, S(a_i)], & ..., & [h_{j,k}, S(a_{|\mathcal{A}|})] \\ ..., & ..., & ..., & ..., & ... \\ [h_{j,|\mathcal{A}|}, S(a_1)], & ..., & [h_{j,|\mathcal{A}|}, S(a_i)], & ..., & [h_{j,|\mathcal{A}|}, S(a_{|\mathcal{A}|})] \end{vmatrix}$$

Fig. 5.2 Power system state for $h_{j,k}$ at Δt

the reader may refer to standard texts on distributed leader election algorithms for more information [24]. The report time intervals Δt are such that $\Delta t \in T_{op}$. The *Coordinator* stores information about the general state of the grid over Δt, as a three dimensional matrix $|\mathscr{MAX}| \times |d| \times N$ as illustrated in Fig. 5.3. We now discuss the structure of the communication network which is necessary to facilitate transferring the data between the devices on the power network.

$$\begin{vmatrix} [\mathcal{SM}(1), h_{1,1}, S(a_{1,...,|\mathcal{A}|})], & ..., & [\mathcal{SM}(1), h_{1,k}, S(a_{1,...,|\mathcal{A}|})], & ..., & [\mathcal{SM}(1), h_{1,|d|}, S(a_{1,...,|\mathcal{A}|})] \\ ..., & & ..., ..., & & ..., ... \\ [\mathcal{SM}(j), h_{j,1}, S(a_{1,...,|\mathcal{A}|})], & ..., & [\mathcal{SM}(j), h_{j,k}, S(a_{1,...,|\mathcal{A}|})], & ..., & [\mathcal{SM}(j), h_{j,|d|}, S(a_{1,...,|\mathcal{A}|})] \\ ..., & & ..., ..., & & ..., ... \\ [\mathcal{SM}(N), h_{N,1}, S(a_{1,...,|\mathcal{A}|})], & ..., & [\mathcal{SM}(N), h_{N,k}, S(a_{1,...,|\mathcal{A}|})], & ..., & [\mathcal{SM}(N), h_{N,|d|}, S(a_{1,...,|\mathcal{A}|})] \end{vmatrix}$$

Fig. 5.3 Power system state—overall grid

5.3.3 *Communication Network*

To communicate power demand requests, billing, and generation information, we support the power network with a communication network built on low cost computational devices. The communication network is comprised of three core sub-networks that are inter-dependent and form a multi-layer radio network. The sub-networks are namely, the *Home Area Network (HAN)*, *Neighbourhood Network (NEN)*, and *Micro-Grid Network (MGN)*. The entire network operates as an asynchronous distributed system with unreliable communication channels and untrustworthy nodes. Therefore, nodes can report false values either due to inherent faults or due to deliberate malicious manipulations.

The HAN is represented by a household $h_{i,i}$ in a cluster of households c_i. The sensors $S(a_i)$ supporting the household appliances are organised to form a wireless sensor network whose communications are coordinated via a wireless sensor network communication protocol such as Bluetooth, ZigBee, and WiFi. This is because communication range between the household devices is a distance of roughly 10–100 m (line-of-sight). The HAN communicates both generated and consumed power to the NEN via the household's $\mathscr{M}$.

The NEN sits outside the home area network, and consists of the cluster of homes c_i that are linked to the metering unit (shared smart meter), $SM(c_i)$ where $SM(c_i)$ is the ith smart meter or also the smart meter associated with c_i which is the ith cluster in the micro-grid. All the SM are connected via a mesh network and are supported by wireless communication protocols such as ZigBee, WiFi, and PLC which offer longer communication ranges and are not impeded by obstructions (i.e. have line-of-sight communication).

Finally the MGN is the point where all of the information from the HAN and NEN is aggregated. We refer to this point as the micro-grid control centre (e.g. coordinating smart meter or utility provider—UP) and use communication protocols, such as WiMAX, Cognitive Radio (CR) or 4G together with a wireless mesh topology to support data transmission from the NEN. The power generation and storage units attached to the UP are modelled to serve as communication points that the UP interacts with to handle demand on the grid. For instance, the UP interacts with the storage unit to either satisfy energy demands on the grid or to store energy that has been purchased from users in u_g.

HAN to NEN and NEN to MGN communications are bilateral so, the communication network controls the power network, conveying messages reporting power generation/consumption from users to the UP and state estimation information is conveyed from the UP (Coordinator) to the users. This is important in providing feedback on all levels of the grid thereby contributing to ensuring trust and performance reliability of the micro-grid. We now discuss how the control network handles power consumption, generation, and state estimation information to ensure causality in operations between the distributed network components.

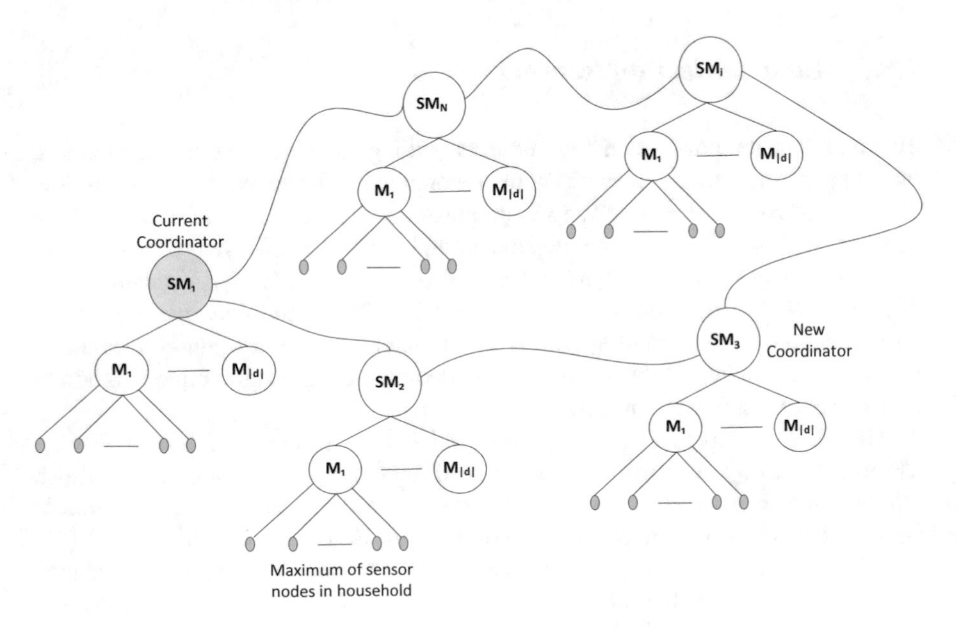

Fig. 5.4 Smart micro-grid schema

5.3.4 *Control Network*

The control network facilitates operation automation of the micro-grid. Building on the communication network, the control network is modelled as an acyclic graph $G = (SM, E)$, where SM is a finite set of smart meters of cardinality N and E is a finite set of edges or communication paths between the smart meters. A SM roots a tree comprised of household grouping devices $\mathscr{M}$, and sensors $S(a_i)$. The maximum of descendent nodes (grouping devices $\mathscr{M}$) a SM can have is bounded by $|d|$ to account for the physical capacity limitations of the SM and to ensure optimal reporting performance. Each $\mathscr{M}$ has a maximum nodal degree of $|\mathscr{M}\mathscr{A}\mathscr{X}|$, which is the maximum number of appliances a household can link to $\mathscr{M}$. This is aimed at modelling a realistic scenario where for efficiency and monitoring purposes, only a certain maximum number of sensor nodes can be connected to a $\mathscr{M}$. Each household, is represented on the tree schema by $\mathscr{M}$. The schema for grid connectivity from the control network perspective, is illustrated in Fig. 5.4. Periodically, the set of SM nodes, come to a consensus (e.g. via a leader election algorithm) on the Coordinator node. Once the coordinator node is agreed on the previous Coordinator node will transfer all existing data to the new Coordinator node. The new Coordinator node will then initiate a power consumption and/or generation collection process requiring all the remaining $N - 1$, SMs to transmit their current readings to the coordinator node. In order to do this, the each SM will communicate to all the $\mathscr{M}$ linked to the SM, a request for data collection and the $\mathscr{M}$ on reception of this message will contact the $S(a_i)$ to collect data. Once the data has been aggregated on all the $\mathscr{M}$, the $\mathscr{M}$ will transmit the values to the SM for subsequent transmission to the Coordinator node (UP).

5.4 Micro-Grid Operation

As mentioned in the previous section, once all the existing data to the new Coordinator node. The new Coordinator node will then initiate a power consumption and/or generation collection process requiring all the remaining $N - 1$, SMs to transmit their current readings to the coordinator node. To do this, the each SM will communicate to all the $\mathscr{M}$ linked to the SM, a request for data collection and the $\mathscr{M}$ on reception of this message will contact the $S(a_i)$ to collect data. Once the data has been aggregated on all the $\mathscr{M}$, the $\mathscr{M}$ will transmit the values to the SM for subsequent transmission to the Coordinator node. The Coordinator node employs a feedback control loop to analyse the data received in order to make billing, scheduling, prediction, and resource allocation (demand management) decisions.

5.4.1 The Feedback Control Loop

The feedback control loop has four main components for data collection and decision making that work with a reference knowledge-base (KB) to perform Monitoring, Analysing, Prediction, and Execution tasks on the power consumption and/or generation data collected. The monitoring component collects/receives data from sensor nodes associated with electrical appliances on the communication network. The monitoring component processes this information to make metering and billing decisions, and transmits relevant information to the Analysis component to handle demand management decisions. Once the demand management decisions have been made, the prediction component takes over to evaluate and forecast future grid behaviour based on current demand requests. Finally, the execution component handles scheduling power distribution according to a prioritisation scheme geared at ensuring gird stability. The information is transmitted back to the sensors and for execution on the power network level. All four components communicate in addition, with the KB to retrieve historical information to support decision making. In the following sections we describe how the components of the control loop work to collect and analyse data in order to ensure efficient grid operation.

5.4.2 Metering

Metering is either manual or automatic. In manual metering, the user in-charge of handling the grouping device manually provides appliance consumption reports by declaring explicitly when and what appliances are to be turned on and the duration of use of the appliances. At this stage, explicit declarations regarding the type of appliance that is being switched on/off must be provided. Depending on the micro-grid demand management system, the user might have to also declare, the projected duration of use of the appliance. Finally, the user declares when the appliance is switched on/off. These values are communicated by the grouping device to the metering device, where the consumption of the household is computed based on the following load type classifications such as resistive, inductive, non-linear, and composite loads [4, 39].

We compute the consumption based on the manual reported values as follows:

$$E = Pt \ldots \tag{5.2}$$

where E is the energy consumption of the appliance during the usage period t, P is the power consumed during t, and $t = T_{OFF} - T_{ON}$. The power consumed is an approximation computed on the basis of the type of appliance used and the load categorisation. The type of appliance A and the time of the day T_{day} is then used to compute the cost of consumption C as follows:

$$C = E * C_t \ldots (2) \tag{5.3}$$

where C_t is the cost of consumption per unit of electricity at T_{day}.

With automated power consumption metering, we assume that each electrical appliance in the household is mapped to a sensor and that periodically the sensor reports the power consumption state to the grouping device $\mathcal{M}$. In order to handle automated data collection, the grouping device employs a data aggregation algorithm that works by requiring every sensor node to report its consumption values for the period t to $\mathcal{M}$. Before we proceed to describe the operation of the algorithms, we first of all provide some basic notation regarding the structure of the sensor network of household appliances.

Each household has a maximum of $|\mathcal{MAX}|$ appliances, each tied to a sensor and that schematically the sensor to mobile (grouping) device relationship is modelled using a subtree where the sensors are the leaf nodes and the grouping device, the parent node. For a more realistic communication model, we expand this description to enable to organise the sensors as an undirected communication graph denoted $G = (S, E)$, where S is the set of the sensor nodes in the household network of appliances and E is the set of all the communication links between the sensor nodes. We say that two sensor nodes s_i and s_j are connected and adjacent, if and only if s_i has a direct uninterrupted line of communication with s_j. The set of adjacent nodes for s_i is given by $N(s_i)$ which implies the set of nodes that are adjacent to the ith sensor.

Reliably computing household power consumption and/or generation over an unreliable sensor network is in fact similar to problem of computing the global state of a distributed system [4, 8]. The snapshot of power consumption and/or generation in the household is a representation of a consistent global state of the local energy consumption and/or generation states of the household's appliances. The global state of a distributed system is obtained via a snapshot modelled as $\{\cup_i (s_i), \cup_{i,j} (e_{i,j})\}$ where (s_i) is the local state of the sensor node s_i and $(e_{i,j})$ is the state of the communication path between s_i and s_j

The distributed snapshot algorithm collects household appliance consumption and/or generation data information over a pre-defined period $[T_a, T_b]$ where T_a denotes the start time and T_b the end time. Due to the computational limitations of the network, instead of collecting consumption and/or generation data in real-time we record snapshots of consumption and/or generation at regular time intervals. This "delayed snapshot" algorithm records the data in the form of a vector of data representing the consumption and/or generation values of the appliances in the HAN recorded over the period t [4]. The vector of consumption and/or generation data is denoted as p_a where $1 \leq a \leq n$ such that a is the number of appliances in the HAN and n is the upper bound on the number of appliances in the HAN. The advantage of using a "delayed snapshot" algorithm is that it enables the collection of data even when there are missing values and outdated data can be collected to obtain a lower bound estimate on current household power consumption patterns.

The "delayed snapshot" algorithm has three steps that are executed atomically [4]. First, the grouping node broadcasts a message to the sensor nodes in the HAN to coordinate data collection. To avoid creating a high overhead, the grouping node employs a minimum spanning tree algorithm to coordinate the transmission and reception of messages to and from the HAN. The spanning tree of G is a subgraph

$T = (V, E)$ that includes vertices from G and that is rooted at the grouping node S_0 (sink node). In addition, the subgraph T has a cluster head S_q that is connected to all the leaf nodes s_i. The cluster head is linked directly by an edge to the grouping node. We ensure resilience to link failure by modelling T as a ρ edge-disjoint spanning tree. Resilience to link failure is achieved by sending ρ copies of the message on the ρ edge-disjoint paths between every pair of nodes that enable a destination node to be reachable despite failure of $\rho - 1$ edges. ρ is considered relatively small i.e. $2 \leq \rho \leq 3$. The edge-disjoint spanning tree can be constructed using standard algorithms for generating a minimum spanning tree [10].

Second, the algorithm uses a control message termed a **MARKER** together with colouring techniques (Black and White), [4, 8] to initiate the collection of consumption and/or generation data. Initially, every node is coloured *White*. As mentioned before, the grouping node initiates the snapshot by first changing its node colour to *Black* and broadcasting a **MARKER** to the nodes in T. Each node in T sets its colour to *Black* upon reception of the **MARKER** message, records the current consumption and/or generation state of the appliance to which it is tagged and resets its colour to *White*. The **MARKER** message contains the sender node identifier and a value to identify the snapshot.

Third, the local cluster and global snapshot are computed. We say the T is formed from a combination of clusters of subtrees such that each subtree is rooted at a cluster node S_q. On reception of the p_a values from the sensor nodes in its subtree, S_q will compute an aggregated value $\sum_{s_i \in S_q}(p_a)$ of power consumption and/or generation for the nodes s_i to which it is linked, and transmit this value to the grouping node. Upon reception of messages from all the S_q the grouping node will compute $\sum \sum_{s_i \in S_q}(p_a)$ the sum of the values obtained from the S_q. The computed value is transmitted to the metering device for billing computation.

5.4.3 *Power Demand Management*

Demand management follows on metering, because we need mechanisms to distribute power fairly and bill users accordingly. In demand management we are therefore faced with two cases. In the first case, all or most of the appliances require manual metering and in the second case, consumption reporting is automated. In the manual case, the user declares the appliances he/she has in the household by linking the appliances to the household grouping device. The user declares the times and duration of appliances use. Demand declarations are thus either manual or automated input from agents on appliances. The projected consumption information is aimed at telling the utility provider how much power the household will be consuming over a projected period. The reason for this approach to data collection on power consumption forecasting is to take into account reporting constraints on the grid where old appliances (without intelligent capabilities) may sometimes be still in use. In these cases, we will have to rely on users to provide input data that can be used together with the appliances' categorisations to compute metering values. On the other hand to account for technological advancements, we also enable automated reports. We begin by describing the projected consumption for a household as a schema for collecting projected consumption data on a per appliance basis. A household's projected consumption (PC) data is computed as a summation of the product of the forecasted per appliance consumption data, $|\mathscr{A}_i|$, where $i = 1 \ldots |\mathscr{A}|$ and $|\mathscr{A}| \leq |\mathscr{M}\mathscr{A}\mathscr{X}|$ represents the total number of appliances in the household, and $\mathscr{T}_i$, the anticipated duration of use of the appliance $|\mathscr{A}_i|$. We compute PC for a given household as follows,

$$PC = \sum_{i=1}^{|A|} \mathscr{A}_i * \mathscr{T}_i\,[a, b] \tag{5.4}$$

To manage demand, the utility station periodically sends out a message to the smart meters to collect forecasted consumption information for the households connected to the smart meter. On reception of this message, the smart meters signal the grouping devices to collect and transmit (PC) information for the household. Each of the devices $\mathscr{M}$ transmits an aggregated PC value to the SM composed from the declared appliance consumption value, $\mathscr{A}_i$ using Eq. (5.2). We handle demand management with a resource allocation algorithm that is constrained by fairness and reliability, and is inspired by the banker's algorithm for resource allocation [15, 30]. Let N be the number of household clusters and LC the number of load categories. In addition, we have the following:

1. A vector of length LC such that, Available $[j] = k$ implies that k units of power type j are available for distribution.
2. An $N \times$ LC matrix such that Total $[i, j] = k$, implies that cluster C_i may request a total k units of power for appliances of type LC_j during a time interval T.
3. An $N \times$ LC matrix where Allocation $[i, j] = k$, implies that C_i has currently been allocated k units of power for appliances of type LC_j
4. An $N \times$ LC matrix such that Need $[i, j] = k$, indicates that C_i needs k units of power more to run appliances of type LC_j.
5. A vector of length LC such that Generated $[j] = k$ indicates that k additional units of power have been generated and can be used for appliances of type LC_j

The Coordinator (UP) node, determines how to distribute power amongst the SM. The SMs, on reception of allocated power units, employ a similar algorithm to distribute power to the households and similarly the $\mathscr{M}$ for the appliances in the household. For simplicity we provide the schema from the perspective of the Coordinator node. We compute the Need matrix as follows:

$$\text{Need}\,[i, j] = \text{Total}\,[i, j] - \text{Allocation}\,[i, j] \tag{5.5}$$

and the power availability vector as follows:

$$\text{Available}\,[j] = \text{Available}\,[j] + \text{Generated}\,[j] \tag{5.6}$$

Allocations are handled by a queue data structure and are processed on a first-come first served basis. On reception of each household's power consumption requests, the UP builds a need matrix for the grid. The UP must then determine whether or not the power requests can be satisfied given the current state of available power on the grid. In order to do so, the UP first assumes that the request can be satisfied and expresses this by assigning a value of "TRUE" to $Satisfiable[i, j]$ to indicate that the amount of power requested by household i for power category j can be provided. The next, the UP performs a test to determine whether or not grid stability would be maintained should the request be granted. The UP does this by performing a test computation in which the requested load is temporarily granted, and the matrices $Available$, $Allocation$, and $Need$ adjusted to reflect the impact of the allocation. A safety check is run to determine whether or not the allocations maintain grid safety (i.e. will ensure that the grid does not break down because it is unable to satisfy power demands).

5.4.4 *Power Demand Forecasting*

Demand forecasting ensures that power demands can be satisfied by encouraging users to make power usage requests at low-peak as opposed to high peak periods by lowering the per unit power price during low peak periods, and incentivizes users with generation units to sell unused power to the utility provider for distribution to units (households or firms) requesting power. The forecasting algorithm is supported by a Kalman filter. Suppose that the observed power consumption data is represented as a vector of multiple time series data $\{y_t\}_{t=1}^{T}$ observed by agents reporting power consumption, and $\{z_t\}_{t=1}^{T}$ represents the vector of system state variables which we represent as discrete time intervals $t = 1, 2, \ldots T$. y_t and z_t are both driven by a stochastic process. We represent our demand forecasting model as a linear state space model of the following form:

$$y_t = Hz_t + v_t \tag{5.7}$$

$$z_t = Bz_{t-1} + \omega_t \tag{5.8}$$

Equation (5.8) describes the relation between the observe time series y_t and the state z_t, assuming that y_t is measured with error that is modelled as a Gaussian error term $v_t \sim N(O, \Sigma_v)$; and H is an $n \times T$ matrix representing the state of consumption of n households over T time intervals.

Equation (5.10) is the transition equation and describes the evolution of the state variables as being driven by the stochastic process of incremental changes represented by ω_t which follows a normal distribution, so $\omega_t \sim N(0, \Sigma_\omega)$. B is a vector representing the available power on the system over T time intervals. We note that ω_t is the amount of new power generated at t. In handling demand forecasting, we must estimate $H, B, \Sigma_v, \Sigma_\omega$ using likelihood based inference with the Kalman filter. Suppose we have reasonable (but not necessarily true) values for Eqs. (5.8) and (5.10) and that these values are equal to

$$\delta = \left\{H^*, B^*, \Sigma_v^*, \Sigma_\omega^*\right\}. \tag{5.9}$$

Let the sample density (or likelihood) function associated with the state space model for the parameters δ be denoted by $f\,(y_1, y_2, \ldots, y_T; \delta)$. By Bayes theorem, we can factor the likelihood function as:

$$f\,(y_1, y_2, \ldots, y_T; \delta) = \prod_{t=1}^{T} f\left(y_t|y^{t-1}; \delta\right) \tag{5.10}$$

where $y^0 = \emptyset$ and $y^{t-1} = (y_1, y_2, \ldots, y_{t-1})$ for $t \geq 2$. To construct the likelihood function we need to derive the densities:

$$f\left(y_t|y^{t-1}; \delta\right), t = 1, 2, \ldots, T \tag{5.11}$$

Since the system is linear and errors are assured to be Gaussian we can do this using the Kalman filter. The Kalman filter is a recursive procedure involving four steps namely 'Initialization', 'Prediction', 'Correction', and 'Likelihood'.

Let $X_{t|s}$ denote the prediction of the consumption load X at time t and that is conditional on information from state estimations available at time s. The Kalman filter is initialized by deriving the best predictor of the initial state $z_{0|0}$ and an estimate of the covariance matrix, $\Sigma_{0|0}^z = E\left[\left(z_0 - z_{0|0}\right)\left(z_0 - z_{0|0}\right)'\right]$. Building on the steady state of the system, we set $z_{0|0} = z^*$ and $\Sigma_{0|0}^z = \Sigma^*; \forall z^* = Bz^*$ and $\Sigma^* = B\Sigma^* B' + \Sigma_\omega$. We set $t = 1$ $\forall z_{t-1|t-1} = z_{0|0}$ and $\Sigma_{t-1|t-1}^* = \Sigma_{0|0}^z$.

At time t, we use $z_{t-1|t-1}$ and $\Sigma_{t-1}^z t - 1$ together with Eq. (5.8) to compute

$$z_{t|t-1} = Bz_{t-1|t-1} \tag{5.12}$$

$$\Sigma_{t|t-1}^z = B\Sigma_{t-1|t-1}^z B' + \Sigma_\omega \tag{5.13}$$

we can then use $z_{t|t-1}$ to construct the forecast $y_{t|t-1} = Hz_{t|t-1}$. having observed y_t, we can construct the forecast error:

$$u_t = y_t - y_{t|t-1} = y_t - Hz_{t|t-1} = v_t + H(z_t - z_{t|t-1}) \tag{5.14}$$

Due to Gaussian errors it follows that $u_t \sim N(0, \Sigma_v + H\Sigma_{t|t-1}^z H')$ and since $y_t = u_t + y_{t|t-1}$, it follows that $f(y_t|y^{t-1}; \delta) = f(u_t; \delta)$. So given $z_{t-1|t-1}$ and $\Sigma_{t-1|t-1}^z$ we can compute $f(y_t|y^{t-1}; \delta)$ from the normal density function and compute $f(y_{t+1}|y^t; \delta)$ with $z_{t|t}$ and $\Sigma_{t|t}^z$ to correct our predictions using the new power consumption and generated data at time t, with the measurements y_t. We correct our predictions by computing

$$z_{t|t} = z_{t|t-1} + K_t(y_t - y_{t|t-1}) = z_{t|t-1} + K_t(y_t - Hz_{t|t-1}) \tag{5.15}$$

$$\Sigma^t_{t|t} = \Sigma^z_{t|t-1} H'(H\Sigma_{t|t-1}H' + \Sigma_v)^{-1} \tag{5.16}$$

The corrected prediction is a linear combination between the old prediction $z_{t|t-1}$ and the current error $y_t - y_{t|t-1}$. Given the linear form, K is chosen such that it minimises the prediction error variance. While $t \neq T$, we increase t and then return to the prediction step. Otherwise we continue and construct the likelihood as follows:

$$L(y^T; \delta) = \prod_{t=1}^{T} f(y_t|y^{t-1}; \delta) \tag{5.17}$$

5.4.5 *Power Demand Scheduling*

Demand scheduling is based on prior information obtained from both the demand management and forecasting phases. This enables the coordinating node estimate grid consumption and generation, that is used to create a power use schedule for appliances'. This encourages users to take advantage of low-demand periods with relatively high levels of available power at a comparatively cheap rate. For example, a household owner might be motivated to turn on the boiler at 2 pm when the per kilowatt rates are relatively low, rather than at 8 pm when the competition for warm water is relatively high and the power costs are higher. However, we note that estimation errors and variability of renewable resources make deviations between the predetermined power supply and demand a necessary consideration.

Since the communication network is supported by low cost inaccurate sensors, measurements are likely to be unreliable. Consequently, errors in reported consumption values may not be intentional attempts to thwart the system. In designing the scheduling mechanism, we address this issue by assuming that appliance power consumption is bounded by a minimum and a maximum value denoted by P_{min} and P_{max} respectively. For each time interval T, the household power consumption vector represented by m_i contains power consumption values for all the household appliances. The corresponding set of power consumption schedules during T is denoted by $P_h = p_{a,h}, \forall a \in \mathscr{A}_h$ and that a $|\mathscr{A}_h| \times T$ matrix represents all the power schedules over the period T. As well, P_h is the sum of the four types of power loads namely, resistive, inductive, non-linear and composite.

The *Interruptibility* and *Consumer Preferences* of devices to be scheduled is considered so, interruptible loads can be stopped and resumed at a later stage, while non-interruptible loads require that the appliance operation executes to termination. We handle interruptible and non-interruptible loads by dividing the execution period of an appliance into several sequential energy periods, where an energy period is an uninterruptible sub-task if the appliance's operation during which pre-specified power consumption is used. The consumer can specify a preference that each appliance be run at a specific time interval. So, each household can negotiate energy needs with the UP, and agree on a schedule that ensures power availability and at a suitable price.

Power consumption scheduling is a constraint based optimisation problem where the global objective function is a summation of objective functions designed to compute the supply cost, household power consumption cost, interruption penalties, energy purchase cost, consumer satisfaction, and storage costs. We express the supply cost model as a combination of the cost of generation and the cost of purchasing energy from the households on the grid. Let p^t denote the price set by the coordinating node at time t, and q_h^t the amount of power purchased for a household h. The total utility cost for electricity $C_u(q_h)$ is given by:

$$C_u(q_h) = \sum_{t=1}^{T} p^t \sum_{h \in H} (g_h^t - l_h^t), \forall g_h^t \geq l_h^t \tag{5.18}$$

where g_h^t denotes the power generated in h during t and l_h^t is the consumption load of h over t. The penalty cost function $C\left(P_{en}^t\right)$ expresses the levy a user has to pay when power scheduling decisions are made over high demand periods. We associate this cost to the duration of interruptions since, interruptible executing device will be pre-empted in favour of the new power demand request. The energy cost of h at t is given by:

$$C_{a,h}^t = \sum_{t \in T} \sum_{a \in A_h} (p^t l_h^t) \tag{5.19}$$

where p^t is a dynamic price provided by the coordinating node. The satisfaction cost function $U\left(p_{a,h}^t\right)$ is represented as follows:

$$U\left(p_{a,h}^t\right) = \begin{cases} \sum_{t \in T} U\left(p_{a,h}^t\right) & \text{if interruptible,} \\ U\left(\sum_{t \in T} p_{a,h}^t \delta t\right) & \text{if deferrable} \end{cases} \tag{5.20}$$

Finally the total cost of power consumption and/or generation per household is the $C\left(P_{en}^t\right)$ and the summation of the $U\left(p_{a,h}^t\right)$ over the period T for all the appliances in h plus the cost of consumption and/or purchase (billed amount) [39]. Finally our power consumption scheduling problem is as follows:

$$min_{q_h, p_h} \left\{ C_u^t(q_h) + \sum_{t \in T} \left(\sum_{a \in A_h} \sum_{h \in H} C_{U,h}^t(p_h) \right) \right\} \tag{5.21}$$

5.5 Adversarial Models

We now study possible attacks, that are triggered to destabilise the grid by either reporting false information to grid users or to prevent communications from occurring in which case a denial of service attack is provoked. Distinguishing legitimate failures from adversarial behaviour is important in preventing subversive behaviour. We focus on attacks that mimic network failure to enable energy theft via cases of non-payment, under-payment or over-payment. Energy theft basically involves a user reporting false consumption amounts or manipulating devices in the household to ensure that the consumed power is partially reported or not at all.

We consider both single and multiple adversaries acting either independently or in collusion. Adversaries can operate from either a fixed location or from multiple locations and also on a temporal basis. Users are classified as either belonging to the smart micro-grid network in which case the user is considered to be internal to the grid, or as not belonging to the grid in which case we say that the user is external to the grid. Attacks can be provoked by both sets of users as individually or as a group. We discuss attacks from the perspective of software manipulations rather than from the brute-force perspective of for example vandalism or physical power line reconnections or manipulations. Since external users wanting to provoke subversive activities will typically be aiming to provoke power theft attacks to result in a breakdown of the grid, we have decided to focus on attacks that are internal to the grid. With internal attacks, the goal of the attacker is merely to avoid paying for the cost of use of electricity and not necessarily to destroy the grid. This is an interesting consideration because as we have mentioned before the stability of the grid relies on user trust in the assurances of reliability and fair billing policies of the grid.

Building on the preceding sections, we suppose we have a small smart micro-grid of size $|dN|$ households, where d is the maximal nodal degree of the smart meters in the network and N the maximum number of smart meters. For a given time interval T, one of the smart meters is elected (by consensus) the leader/coordinator node and this node handles all of the billing and demand management information. Let h' denote the set of grid households whose power consumption reports have been compromised, where $h' \subseteq H$ and H denotes the set of all households on the grid consuming power. As mentioned before, the consumption values of each household are mapped onto a vector of length $|\mathscr{M}\mathscr{A}\mathscr{X}|$ where each vector entry $\{i \ \forall a_i \in \mathscr{A}\}$ represents the value reported for an appliance a_i at a time t.

5.5.1 False Data Injection Attacks

False data injection attacks can be provoked by intercepting and modifying the vector of consumption values transmitted to the smart meter. For instance, forecasted values can be modified to distort household power scheduling requests and yet bill the households as though the consumption actually occurred. We consider two false data injection attack scenarios as follows:

5.5.1.1 Scenario 1

The adversary h_{adv}, starts by declaring that consumption and/or generation reports from his/her household appliances can only be made manually because the devices are old and not equipped with sensors. In this case, suppose the "correct" consumption report vector is denoted h where $a_i \quad \forall 1 \leq i \leq |\mathscr{M}\mathscr{A}\mathscr{X}|$ represents the state estimation values in $h_{i,i}$; and that the modified vector h_{adv} represents the consumption report vector that $h_{i,i}$ transmits. In forming h_{adv}, $h_{i,i}$ aims to compute the combined sum of energy across the appliances in the household and to spread the consumption over the all the a_i. So, appliances requiring more power appear to be consuming less and ones requiring less power appear to be consuming more. Since metering schemes charge per power type, this results in an overall drop in the final amount billed (Fig. 5.5).

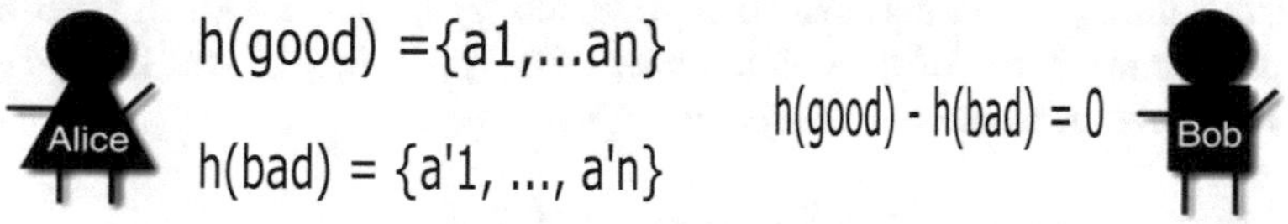

Fig. 5.5 False data injection attack: example

5.5.1.2 Scenario 2

The adversary household h_{adv} wishes to distort the consumption report vectors of a group of households h'. This attack type can be classified as a case of misattribution which is a variant of a power theft attack where a user tricks the system into attributing his/her power consumption values to some other user(s) on the system. The goal of the adversary is to escape detection by spreading his/her consumption over the compromised nodes to ensure that the compromised values each reflect a fraction of the values reported from h_{adv} so as to reduce the chance of discovery. This is done determining the fraction of households required to minimise the risk of detection where the fraction of compromised households is given by $\frac{|h'|}{dN}$. Assuming that h_{adv} has a set of appliances a_i, then for each $h \in h'$ the amount to be added to the consumption of each $h \in h'$ is computed as follows:

$$h = \frac{\sum_{i=1}^{|\mathscr{M}\mathscr{A}\mathscr{X}|} a_i \times dN}{|h'| \times |\mathscr{M}\mathscr{A}\mathscr{X}|} \tag{5.22}$$

Once h_{adv} has decided on the amounts of power consumption values to be shifted off to the households h', h_{adv} must proceed to find a way to compromise a sufficient set of nodes. To compromise the required set of nodes, h_{adv} must intercept consumption report signals of at least $|h'|$ households [22]. Assume that there are $|h'|$ metering units that h_{adv} needs to compromise. h_{adv} based on observations of the measurements at the metering units and Eq. (5.22) above, and then injects the malicious measurements into the compromised meters. The injected malicious measurements can provoke state estimation errors without detection because, the changes are designed to mimic or imitate variations due to channel transmission errors which typically reflect only small deviations from a threshold value.

Suppose that the original measurements of $h_{i,i}$ are accepted by the coordinating node as valid. The distorted measurements $h'_{i,i}$ where $h'_{i,i} = h_{adv} = h \pm \alpha$ can also be accepted as valid, if α is structured to form a linear combination of the column vectors of the consumption report matrix $SM(i)$ such that $\alpha = SM(i)\beta$.

Since, we know that $h_{i,i}$ passes the verification test at the coordinating node, the detection threshold for false measurements is given by $\|h_{i,i} - SM(i)\hat{a}\| \leq \tau$, where τ is the detection threshold and $\hat{a_{bad}}$ is the vector of estimated state variables obtained from h_{adv} and that can be represented $\hat{a} \pm \beta$. If $\alpha = SM(i)\beta$ then a is a linear combination of the column vectors $h_{i,1}, \ldots, h_{i,d}$ of $SM(i)$ and so the L_2-norm of the measurement residual is:

$$\|h_{adv} - SM(i)\hat{a}_{bad}\| = \|h_{i,i} + \alpha - SM(i)\hat{a} + \beta\| \tag{5.23}$$

$$= \|h_{i,i} - SM(i)\hat{a} + (\alpha - SM(i)\beta)\| \tag{5.24}$$

$$= \|h_{i,i} - SM(i)\hat{a}\| \leq \tau \tag{5.25}$$

Therefore, the L_2-norm of the measurement residual of h_{adv} is less than the threshold τ which implies that h_{adv} can also pass the bad measurement detection.

5.5.2 *Denial-of-Service Attacks*

A denial-of-service attack can be provoked by manipulating the power distribution scheme into a situation of deadlock. We consider two cases the first involves a single malicious user/household wanting to gain unfair access to power, while the second case deals with a malicious user/household that colludes with other households to gain unfair access to power.

5.5.2.1 Scenario 1

Suppose a malicious node belonging to the SMG and denoted h_{adv} wishes to provoke a situation of deadlock in order to gain unfair and superior access to power at a low cost. In this case, the goal of h_{adv} is to make all his/her requests during off-peak periods but at the cost of preventing other households from accessing the grid during this period. In order to provoke the attack, h_{adv} will begin by showing good faith and supplying the coordinating node with a certain amount of generated power, say αat a time t_{OP} that falls during the off-peak period. This power α is added to the amount that is currently available and the information is broadcast to the members of the grid. On reception of the information about the amount of power available h_{adv} makes a request for power that is equivalent to the amount available and repeats this procedure every time power is declared available during off-peak period. In this way, assuming h_{adv} has the consumption and/or storage capacity, he/she is able to deny the other grid users access to power at periods when the price is reasonable. Once h_{adv} has acquired the power, h_{adv} then proceeds to offer the power at a lower price to grid users at peak periods, at a lower price than the official going rate (Fig. 5.6).

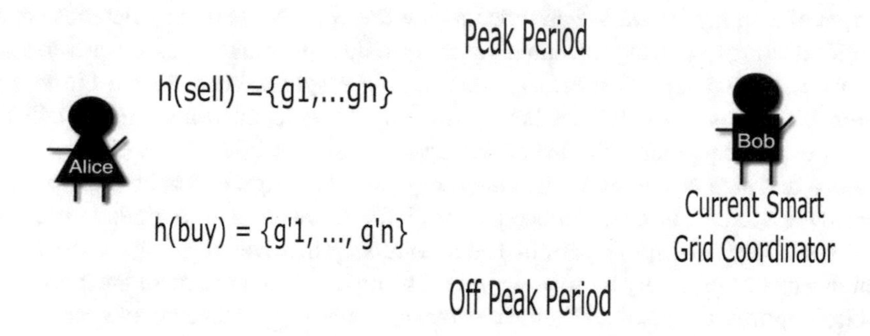

Fig. 5.6 Denial-of-service attack: example

5.5.2.2 Scenario 2

This attack can also be modified to allow colluding and/or coalitions of users to mount power theft (forging) attacks against the SMG. In the case of collusions, a malicious user h_{adv} begins by persuading households in its neighbourhood to join a malicious power request group P_{mal}. Users in P_{mal} make power requests to h_{adv} which are transmitted $\pm\alpha$ a value calculated by h_{adv} to ensure that all the available power is at the coordinating node is depleted. h_{adv} then distributes the power to the nodes in P_{mal} and sells the remaining power at a later time, at a price $p > p_{buy}$, where p_{buy} is the price at which h_{adv} purchased the power but such that $p < p_{actual}$ where p_{actual} is the price offered by the coordinating node at peak period.

5.5.3 Coalition and Collusion Attacks

Finally in the coalition attack we consider cases of cheating where the adversaries are either internal to or external from the SMG. In this case the adversaries are assumed to have only partial knowledge of the topology of the communication network and the power generation/consumption patterns. The goal of the adversaries therefore is to simply modify the data that finally arrives at the SMs. In order to provoke this attack, two malicious users say, h_{adv} and h'_{adv} employ a combination of techniques such as message forging, tampering, and phishing to obtain information to modify the data reported by the devices on the network.

5.5.3.1 Scenario 1

Here, h_{adv} and h'_{adv} agree to form a coalition with the users in their respective clusters c_{adv} and c'_{adv}. The goal here is to cheat the SMG system by employing a variant of a replay attack. We assume that the SMG is able to distinguish false from correct readings by using threshold values to detect invariances, so the adversaries must ensure that reports from c_{adv} and c'_{adv} are spread between the clusters to minimise the risk of detection but at the same time to maximise the benefit for the users who cooperate with the adversaries.

As a first step in mounting the attack h_{adv} and h'_{adv} agree to cooperate to cheat the power system, in order to use power at for instance peak periods at the cost one would pay at off-peak period. The second step involves forming the coalition, which happens primarily by sending out phishing text messages to all the household control/reporting devices $\mathscr{M}$ in the clusters c_{adv} and c'_{adv}. Once an adequate set of malicious users has been formed, h_{adv} and h'_{adv} can now proceed to provoke replay attacks.

The replay attack is provoked by forging the timestamps associated with the measurements reported from $a_1, \ldots, a_{|\mathscr{MAX}|}$ for all the households cooperating with h_{adv} and h'_{adv}, so the old measurements appear to have been newly generated. With the help of a series of permutations and combinations h_{adv} and h'_{adv} reorder the readings from the appliances they control, in order to make the old readings appear to be a new sequence of readings for the new period Δt_{new} (Fig. 5.7).

Suppose $H_{adv} = h_1, \ldots, h_y$, where $y \leq dN$, is the set of users who agree to participate in the coalition of users provoking the replay attack and $\mathscr{L}_{H_{adv}}$ the consumption of all $h_i \in H_{adv}$ during Δt, where $\Delta t \in \mathscr{T}$. All households h_i record a series of measurements over various reporting intervals which h_{adv} and h'_{adv} intercept. During the attack interval Δt_{adv}, h_{adv} and h'_{adv} intercept the reports from h_i and modify as follows:

$$L^t_{h_{adv}} = \begin{cases} L^t_{new}, & \text{if } t \notin \Delta t_{adv} \\ L^t_{old}, & \text{if } t \in \Delta t_{adv} \end{cases} \tag{5.26}$$

where $L^t_{h_{adv}}$ is the consumption report for h_{adv} and L^t_{new} is the correct value expected for the consumption during Δt, L^t_{old}transmitted during Δt_{adv} is a permutation of values selected from the different readings obtained in previous time intervals Δt. The combinations are selected using $L^t_{old}(n, r) = \frac{n!}{r!(n-r)!}$ where

h(good) ={a1,...an}
h(bad) = {a'1, ..., a'n}
Alice

h(good) ={a1,...an}
h(bad) = {a'1, ..., a'n}
Eve

Fig. 5.7 Collusion and coalition attacks: example

$n \leq |\mathscr{MAX}|$ and r the number of a_i to be modified for a successful attack. Finally, $L^t_{old}(n, r)$ represents the number of ways in which the reported values can be combined to support the attack successfully without repeatability.

5.5.3.2 Scenario 2

Here h_{adv} and h'_{adv} work together to coordinate random data perturbation attacks supported by a coalition of adversarial households. In this case, the strategy is for the households to modify power consumption and/or generation data reports before they are reported first to the cluster head (smart meter). This can happen in two ways. In the first case, the household declares that reports can only be made manually, and then proceeds to report false values for all or the majority of appliances in the household. In the second case, the modifications are provoked by adding noisy signals to the reported values. We express these additions from the manual and automated point as follows:

$$L_{h_{adv}} = \begin{cases} L_{normal}, & \text{if } t \notin \Delta t_{adv} \\ L_{normal} \pm \rho_{adv}, & \text{if } t \in \Delta t_{adv} \end{cases} \tag{5.27}$$

where $L_{h_{adv}}$ represents the modified value transmitted to SM coordinating the cluster c_{adv} and the same is true for h'_{adv}. ρ_{adv} represents the amount of noise added to the signal to reduce the reported value to one that maximises the advantage to the adversary while minimising the risk of disclosure.

5.6 Conclusions and Future Work

5.6.1 *Summary*

In this chapter, we presented a reference model for a SMG designed to operate in a resource constrained environment optimally. We use the term *resource constrained* to describe a power network constructed or based on environmentally-friendly renewable energy sources coordinated by distributed algorithms running over a lossy network composed of unreliable hardware such as sensors and rudimentary mobile devices communicating over wireless networks. We began by proposing a general framework for the structure of the SMG showing how three network layers namely, the power, communication, and control networks interact to coordinate power supply and demand management. As a follow up we considered the control network in more depth providing supporting algorithms and then followed that with a discussion of likely attack models aimed at power theft. The reason for focusing on power theft is that grid users have thee incentive of a reliable power supply source and so are not likely to want to do things to directly provoke a breakdown of the grid. However, power theft allows users who are able to do this to gain a little bit more unfair advantage over other grid users. The algorithms proposed to support the framework were analysed from the performance perspective and demonstrated to be feasible and operate correctly under the constraints of the environment.

In addition, we presented a control network architecture to handle power consumption and/or generation data transmissions. We considered the fact that the control network could be modelled to operate as a feedback control loop composed of monitoring, analysing, prediction, and execution modules that interact with the devices in a household. To support the feedback control loop, some algorithms were proposed to handle operations such as metering, power demand requests, forecasting, and scheduling. We studied these algorithms as a means of ensuring dependable and reliable micro-grid operation over an unreliable communication network.

5.6.2 *Future Work*

As future work, we are currently working on modelling the algorithms required to support the control network as well as attack mitigation algorithms. Our next step, will be implementing a prototype simulation environment where we can test performance and the robustness of our attack mitigation algorithms. Possible avenues for future work would include developing a simulation environment for a resource constrained smart micro-grid where the control network algorithms are integrated.

References

1. A. Abur and A. Gómez Expósito. *Power System State Estimation: Theory and Implementation*. CRC Press, Boca Raton, FL, USA, 2004.
2. P. L. Ambassa, A. Kayem, S. D. Wolthusen, and C. Meinel. Secure and reliable power consumption monitoring in untrustworthy micro-grids. In Robin Doss, Selwyn Piramuthu, and Wei ZHOU, editors, *Future Network Systems and Security*, volume 523 of *Communications in Computer and Information Science*, pages 166–180. Springer International Publishing, Springer International Publishing Switzerland, 2015.
3. P. L. Ambassa, A. Kayem, S. D. Wolthusen, and C. Meinel. Privacy violations in constrained micro-grids: Adversarial cases. In *2016 30th International Conference on Advanced Information Networking and Applications Workshops (WAINA)*, pages 601–606, March 2016.
4. P. L. Ambassa, S. D. Wolthusen, A. Kayem, and C. Meinel. Robust snapshot algorithm for power consumption monitoring in computationally constrained micro-grids. In *Smart Grid Technologies - Asia (ISGT ASIA), 2015 IEEE Innovative, Bangkok, Thailand*, pages 1–6, Piscataway, NJ, USA, 3–6 Nov.2015 2015. IEEE Press.
5. A. Baiocco, S. Wolthusen, C. Foglietta, and S. Panzieri. A model for robust distributed hierarchical electric power grid state estimation. In *Innovative Smart Grid Technologies Conference (ISGT), 2014 IEEE PES*, pages 1–5, Piscataway, NJ, USA, Feb 2014. IEEE Press.
6. A. Baiocco and S. D. Wolthusen. Dynamic forced partitioning of robust hierarchical state estimators for power networks. In *Innovative Smart Grid Technologies Conference (ISGT), 2014 IEEE PES*, pages 1–5, Piscataway, NJ, USA, Feb 2014. IEEE Press.
7. P. Buchana and T. S. Ustun. The role of microgrids amp; renewable energy in addressing sub-saharan Africa's current and future energy needs. In *Renewable Energy Congress (IREC), 2015 6th International*, pages 1–6, Sousse, Tunisia, 24–26 March 2015. IEEE Press.
8. K. Mani Chandy and L. Lamport. Distributed snapshots: Determining global states of distributed systems. *ACM Trans. Comput. Syst.*, 3(1):63–75, February 1985.
9. J. Conti, P. Holtberg, J. Diefenderfer, A. LaRose, J. T. Turnure, and L. Westfall. International energy outlook 2016. URL, June 2016.
10. T. H. Cormen, C. Stein, R. L. Rivest, and C. E. Leiserson. *Introduction to Algorithms*. The MIT Press, USA, 3rd edition, 2009.
11. Y. Feng, C. Foglietta, A. Baiocco, S. Panzieri, and S. D. Wolthusen. Malicious false data injection in hierarchical electric power grid state estimation systems. In *Proceedings of the Fourth International Conference on Future Energy Systems*, e-Energy '13, pages 183–192, New York, NY, USA, 2013. ACM.
12. A. Gómez Expósito, A. Abur, A. de la Villa Jaén, and C. Gómez-Quiles. A multilevel state estimation paradigm for smart grids. *Proceedings of the IEEE*, 99(6):952–976, jun 2011.
13. A. Gómez Expósito, A. Abur, A. de la Villa Jaén, C. Gómez-Quiles, P. Rousseaux, and T. van Cutsem. A taxonomy of multilevel state estimation methods. *Electric Power Systems Research*, 81(4):1060–1069, apr 2011.
14. C. Gómez-Quiles, A. de la Villa Jaén, and A. Gómez Expósito. A factorized approach to WLS state estimation. *IEEE Transactions on Power Systems*, 26(3):1724–1732, aug 2011.
15. D. A. Hensgen, D. L. Sims, and D. Charley. A fair banker's algorithm for read and write locks. *Information Processing Letters*, 48(3):131–137, 1993.
16. K. Iniewski. *Smart Grid: Infrastructure and Networking*. McGraw Hill, New York, NY, USA, 2013.
17. A. Kayem, C. Meinel, and S. D. Wolthusen. A smart micro-grid architecture for resource constrained environments. In *2017 IEEE 31st International Conference on Advanced Information Networking and Applications (AINA)*, pages 857–864, March 2017.
18. M. Klonowski and D. Pajak. Electing a leader in wireless networks quickly despite jamming. In *Proceedings of the 27th ACM Symposium on Parallelism in Algorithms and Architectures*, SPAA '15, pages 304–312, New York, NY, USA, 2015. ACM.

19. G. N. Korres. A distributed multiarea state estimation. *IEEE Transactions on Power Systems*, 26(1):73–84, feb 2011.
20. D. Koss, D. Bytschkow, P. K. Gupta, B. Schätz, F. Sellmayr, and S. Bauereiss. Establishing a smart grid node architecture and demonstrator in an office environment using the SOA approach. In *Proceedings of the First International Workshop on Software Engineering Challenges for the Smart Grid*, SE4SG '12, pages 8–14, Piscataway, NJ, USA, 2012. IEEE Press.
21. R. Kuwahata, N. Martensen, T. Ackermann, and S. Teske. The role of microgrids in accelerating energy access. In *3rd IEEE PES Innovative Smart Grid Technologies Europe (ISGT Europe)*, pages 1–9, Piscataway, NJ, USA, Oct 2012. IEEE Press.
22. Y. Liu, P. Ning, and M. K. Reiter. False data injection attacks against state estimation in electric power grids. In *Proceedings of the 16th ACM Conference on Computer and Communications Security*, CCS '09, pages 21–32, New York, NY, USA, 2009. ACM.
23. Z. Liu. Chapter 3 - a global energy outlook. In Zhenya Liu, editor, *Global Energy Interconnection*, pages 91–100. Academic Press, Boston, 2015.
24. N. A. Lynch. *Distributed Algorithms*. Morgan Kaufmann Publishers Inc., San Francisco, CA, USA, 1996.
25. A. M. C. Marufu, A. Kayem, and S. D. Wolthusen. A distributed continuous double auction framework for resource constrained microgrids. In *10th International Conference on Critical Information Infrastructures Security (CRITIS 2015), October 5–7, 2015, Berlin, Germany*, pages 183–196. Vol. 9578, Lecture Notes in computer Science, Springer-Verlag, 2015.
26. A M. C. Marufu, A. Kayem, and S. D. Wolthusen. Fault-tolerant distributed continuous double auctioning on computationally constrained microgrids. In *2nd International Conference on Information systems Security and Privacy (ICISSP 2016), February 19–21, 2016, Rome, Italy*, pages 448–456. SCITEPRESS, 2016.
27. A. M. C. Marufu, A. Kayem, and S. D. Wolthusen. Power auctioning in resource constrained micro-grids: Cases of cheating. In *11th International Conference on Critical Information Infrastructures Security (CRITIS 2016), October 10–12, 2016, Paris, France*, page (To Appear). Lecture Notes in computer Science, Springer-Verlag, 2016.
28. G. M. Mathews. An optimal hierarchical algorithm for factored nonlinear weighted least squares state estimation. In *Proceedings of the 2012, 3rd IEEE PES International Conference on Innovative Smart Grid Technologies (ISGT Europe 2012)*, pages 1–6, Berlin, Germany, oct 2012. IEEE Press.
29. E. D. Moe and A. P. Moe. Off-grid power for small communities with renewable energy sources in rural guatemalan villages. In *Global Humanitarian Technology Conference (GHTC), 2011 IEEE*, pages 11–16, Piscataway, NJ, USA, Oct 2011. IEEE Press.
30. E. Louise Moser and P. M. Melliar-Smith. The world banker's algorithm. *Journal of Parallel and Distributed Computing*, 9(4):369–373, 1990.
31. R. Nagaraj. Renewable energy based small hybrid power system for desalination applications in remote locations. In *2012 IEEE 5th India International Conference on Power Electronics (IICPE)*, pages 1–5, Piscataway, NJ, USA, Dec 2012. IEEE Press.
32. D. N. Nikovski, Z. Wang, A. Esenther, H. Sun, K. Sugiura, T. Muso, and K. Tsuru. Smart meter data analysis for power theft detection. In Petra Perner, editor, *Proceedings of the 9th International Conference on Machine Learning and Data Mining in Pattern Recognition (MLDM 2013)*, volume 7988 of *Lecture Notes in Computer Science*, pages 379–389, New York, NY, USA, jul 2013. Springer-Verlag.
33. M. Perdue and R. Gottschalg. Energy yields of small grid connected photovoltaic system: effects of component reliability and maintenance. *IET Renewable Power Generation*, 9(5):432–437, 2015.
34. F. C. Schweppe and J. Wildes. Power system static-state estimation, part i: Exact model. *IEEE Transactions on Power Apparatus and Systems*, PAS-89(1):120–125, jan 1970.
35. F. C. Schweppe and J. Wildes. Power system static-state estimation, part ii: Approximate model. *IEEE Transactions on Power Apparatus and Systems*, PAS-89(1):125–130, jan 1970.
36. T. B. Smith. Electricity theft: a comparative analysis. *Energy Policy*, 32(18):2067–2076, 2004.

37. T. van Cutsem and M. Ribbens-Pavella. Critical survey of hierarchical methods for state estimation of electric power systems. *IEEE Transactions on Power Apparatus and Systems*, PAS-102(10):3415–3424, oct 1983.
38. Z. Wang and M. Lemmon. Stability analysis of weak rural electrification microgrids with droop-controlled rotational and electronic distributed generators. In *2015 IEEE Power Energy Society General Meeting*, pages 1–5, Piscataway, NJ, USA, July 2015. IEEE Press.
39. G. K. Weldehawaryat, P. L. Ambassa, A. M. C. Marufu, S. D. Wolthusen, and A. Kayem. Secure and decentralized power consumption scheduling in constrained micro-grids. In *2nd Workshop on security of Industrial Control Systems and Cyber Physical Systems (CyberICPS 2016)*, page (To Appear), 2016.
40. T. Winther. Electricity theft as a relational issue: A comparative look at Zanzibar, Tanzania, and the Sunderban Islands, India. *Energy for Sustainable Development*, 16(1):111–119, 2012.

Chapter 6
The Design and Classification of Cheating Attacks on Power Marketing Schemes in Resource Constrained Smart Micro-Grids

Anesu M. C. Marufu, Anne V. D. M. Kayem, and Stephen D. Wolthusen

Abstract In this chapter, we provide a framework to specify how cheating attacks can be conducted successfully on power marketing schemes in resource constrained smart micro-grids. This is an important problem because such cheating attacks can destabilise and in the worst case result in a breakdown of the micro-grid. We consider three aspects, in relation to modelling cheating attacks on power auctioning schemes. First, we aim to specify exactly how in spite of the resource constrained character of the micro-grid, cheating can be conducted successfully. Second, we consider how mitigations can be modelled to prevent cheating, and third, we discuss methods of maintaining grid stability and reliability even in the presence of cheating attacks. We use an Automated-Cheating-Attack (ACA) conception to build a taxonomy of cheating attacks based on the idea of adversarial acquisition of surplus energy. Adversarial acquisitions of surplus energy allow malicious users to pay less for access to more power than the quota allowed for the price paid. The impact on honest users, is the lack of an adequate supply of energy to meet power demand requests. We conclude with a discussion of the performance overhead of provoking, detecting, and mitigating such attacks efficiently.

Keywords Smart micro-grids · Cheating attacks · Power auctioning

A. M. C. Marufu (✉)
Department of Computer Science, University of Cape Town, Cape Town, South Africa
e-mail: amarufu@cs.uct.ac.za

A. V. D. M. Kayem
Hasso-Plattner-Institute, Faculty of Digital Engineering, University of Potsdam, Potsdam, Germany

S. D. Wolthusen
Department of Mathematics and Information Security, Royal Holloway, University of London, Egham, Surrey, UK

Norwegian Information Security Laboratory, Gjovik University College, Norwegian University of Science and Technology, Trondheim, Norway
e-mail: stephen.wolthusen@ntnu.no

A. V. D. M. Kayem et al. (eds.), *Smart Micro-Grid Systems Security and Privacy*, Advances in Information Security 71, https://doi.org/10.1007/978-3-319-91427-5_6

6.1 Introduction

In rural/remote regions it is sometimes reasonable and cost effective to empower a local population with low-cost, resource-constrained information technology to manage power. One plausible method of powering such regions is to use an autonomous smart micro-grid supported by a lossy communication network [1]. Kayem et al.'s reference resource constrained micro-grid architecture incorporates power flow, communication, and control network structures that encapsulates the aforementioned properties. The model describes a remote community-based set-up with clustered households relying mostly on renewable energy resources for power. Such a model, although conceptual was novel and has allowed for some results to be drawn for further questions asked in subsequent work towards scheduling [2]; power consumption monitoring [3, 4]; and power auctioning [5–8].

Smart micro-grids are fragile, requiring fair demand management in addition to generator matching to maximize the overall system utility [5, 9, 10]. Evidently, RC smart micro-grids would benefit grossly from an efficient, reliable and secure power management or resource allocation system. A decentralised Continuous Double Auctioning (CDA) algorithm is one such resource-allocation mechanism that can be employed as an additional power management application [5] since it allows: efficient resource-allocation without the need of a centralised auctioneer (like [11, 12]); local decision-making by multiple buyers and sellers (distributed generators or demand) who have incomplete and imperfect information; ability to be robust in dynamic environments; high efficiency while maintaining low computational cost; and ensures fairness in resource allocation [13]. Efficient resource allocation (hence power management) is an emergent behaviour of the complex interactions of the individual self-interested trading agents, with transactions corresponding to allocations. According to Smith's seminal work [14] a CDA can be described as a market mechanism where high efficiency is achieved by a relatively small number of selfish human traders, in decentralised environment, where no single agent has complete and perfect information about the system. Smith demonstrated that transactional prices converge to the market's theoretical competitive equilibrium price. These results were novel as they showed that markets governed by a decentralised mechanism, such as the CDA, do not require to be large to be efficient, as had previously been assumed. Many subsequent research endeavours in this area have been heavily influenced by this work.

In market-based control, software agents[1] provide truly automated and distributed control systems, rather than relying on a central auctioneer [16] or human intervention. Agent technology can contribute to different aspects of consumer buying—deciding what to buy, whom to buy it from, how much to pay, and the actual trade of goods for money [17]. The degree and the level of sophistication of such an automation can benefit the CDA. To achieve this, software agents that act

[1] Also referred to as a bargaining agents by Priest and Tol [15].

on behalf of human users (as delegates) are required to fulfil the user requirements and expectations [18, 19]. Such an agent must exhibit the following properties: autonomy; adaptivity, pro-activeness; reactivity; prediction; social ability; ability to learn; and sometimes mobility [18]. Thus, the agent requires a strategy for negotiation which is at least as effective as a qualified human being in the same situation [15]. Multi-agent based distributed energy resource management has been studied extensively in [20–22]. In this chapter, we consider trades in an auction occur when trading agents (*TA*s) interact with one another (i.e., they buy and sell goods or services). Each *TA* knows information about itself and can collect information made public from auction market. The *TA* s employ some heuristics in their 'strategy' to handle incomplete information and the dynamic market environment [18]. The input of the strategy includes private information (limit price, eagerness to trade, etc.) and public information (such as outstanding ask/bid, last transaction price, etc.). The output of the strategy is an offer to be submitted. Since the Santa Fe Double Auction Tournament (SFDAT) [23] was conducted, several bidding strategies have been developed to determine the most efficient strategy best for the CDA market. These include: Gode and Sunder's [24] Zero-Intelligence (ZI) strategy; Cliff's [25, 26] Zero-Intelligence Plus (ZIP) strategy; Preist and Tol's [15] CP strategy; Gjerstad and Dickhaut's [27] Gjerstad-Dickhaut (GD) strategy; Tesauro and Das' [28] Modified Gjerstad-Dickhaut (MGD) strategy; Tesauro and Bredin's [29] Extended Gjerstad-Dickhaut (GDX) strategy; He et al.'s [30] Fuzzy Logic (FL) strategy; Vytelingum et al. [31] Risk-Based (RB) strategy; Cliff's [32] ZIP60 strategy; Vytelingum's [19].

The CDA algorithm design and deployment should take into account the fact that actors in the auction market are self-interested [33], implying they may misrepresent their preferences (e.g. amount of electricity required, the capacity they can supply and prices they would accept) or even change agent's bidding strategy in order to maximise their profit [7]. The tendency for participants to cheat or employ strategies to gain some economic advantage disrupts services and hinders trust which is necessary for incentivising energy sharing among the members of such a community. Cheating has been arguably the most significant group of attacks forming the bulk of all internet frauds [34, 35]. Cheating unlike other fraud categories leaves no direct evidence of its occurrence, while financial loss resulting from such cheating behaviour cannot be precisely measured. Some reasons that encourage cheating are: cheap user pseudonyms; greater information asymmetry; lack of personal contact between participants; and the tolerance of bidders [36]. More so, we acknowledge that adversaries are usually attracted to popular platforms as ideal landscapes for exploits; incorporating auction models into the increasingly popular smart micro-grid platform could have serious security consequences on such cyber-critical systems.

Most widely studied classic forms of cheating occur mostly in single-sided auctions and to a less extent centralised CDAs. Marufu et al. [7] argue that cheating is auction mechanism specific; thus, a decentralised CDA is a fairly comprehensive schemes that discourage some standard cheating forms such as multiple bidding, bid shading, rings, shill bidding, false bids, etc. Trevathan [36] analyses the aforementioned standard cheating forms. Marufu et al. [7] further suggest that automation in decentralised CDAs brought about by employing software agents[2] can open up avenues for automated forms of cheating. Thus, cheating is auction mechanism specific; which implies that cheating forms and the related countermeasures are dependent on the auction mechanism [7]. Given the disruptive potential of cheating attacks in auctions, for general use, [36–40], and specifically for auction augmented smart micro-grids [7, 8], it is prudent to design mechanisms for detecting and defending such CDAs against such attacks.

To the best of our knowledge, research in [40] and [41], is the only closest work that specifically addresses CDA security. Wang and Leung in [40], describe an anonymous and secure CDA protocol for electronic marketplaces, which is strategically equivalent to the traditional CDA protocol. Trevathan et al. in [41], demonstrated that, Wang and Leung's scheme [40] allows the identity of a bidder to be revealed immediately after his/her first bid and improved on their scheme. Marufu et al. [7] identify two forms of cheating realised by changing the trading agent (*TA*) strategy of some agents in a homogeneous CDA scheme. In one case, an adversary gains control and degrades other trading agents' strategies to gain more surplus. While in the other, K colluding trading agents employ an automated coordinated approach to changing their *TA* strategies to maximize surplus power gains. In the follow up work, Marufu et al. [8] propose a novel scheme to circumvent power auction cheating attacks. The scheme works by employing an exception handling mechanism that employs cheating detection and resolution algorithms.

This body of CDA security work and attack models, while useful, suffers from the following limitations: (a) There is no general purpose framework that unifies the design of cheating attacks in CDAs; (b) most attack models in the literature do not explicitly account for the functionality and architecture of a CDA and, as a consequence, ignore or overlook the properties that could help advise system defenders. Thus, the main contribution in this chapter is the development of the Automated-Cheating-Attacks (*ACA*) Framework that enables the design of cheating attacks for the development of attack detection methods and tools. We design cheating attacks detailing their feasibility, attack procedure, performance overheads and drawbacks. Using the *ACA* Framework we classify the different type of attacks that result from an attacker who has intentions to cheat. Furthermore present a plausible, resource-aware mitigation solution to deter cheating attacks on decentralised CDA.

The remainder of this paper is structured as follows: In Sect. 6.2 we detail the *ACA* Framework, and subsequent design of cheating attacks in Sect. 6.3. In Sect. 6.4 we categorise automated cheating attacks based on the proposed models. Section 6.5 outlines a plausible counter measure to mitigate the designed cheating attacks. Section 6.6 discuses related work and a summary concludes this book chapter linking the objectives of the chapter to the presented work in Sect. 6.7

---

[2]Also referred to as bargaining agent by Priest and Tol [15].

6.2 Framework Design

As part of our first contribution in this book chapter, we propose a framework that allow unifying of a variety of cheating attack models for CDA. The Automated-Cheating-Attacks (*ACA*) Framework is inspired by the work of Adepu and Mathur [42]. We base our framework on Adepu and Mathur's work as it is simple, allows easy design of attacks and to the best of our knowledge is the closest framework that describes designing of attacks considering with more coverage. By leveraging some aspects of Adepu and Mathur's framework and extending it to capture the CDA and behavioural aspects, we are able to develop an arguably effective framework for designing cheating attacks. Figure 6.1 depicts the *ACA* Framework. We envisage that our framework allows researchers to design a variety of cheating attacks on CDAs in general, and on decentralised CDAs specifically, for the assessment of attack detection methods and tools.

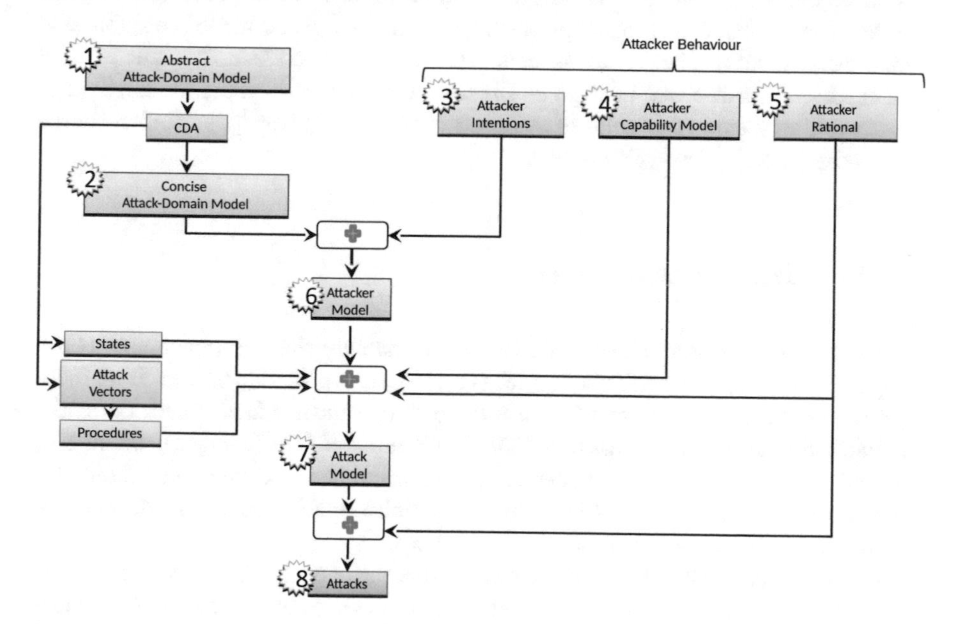

Fig. 6.1 *ACA* Framework for deriving attacks

For simplicity and ease of conceptualising, the *ACA* Framework has the following main components or models, namely: *Attack-Domain* model, *Attacker* model, and *Attack* model, as these directly guide the design of attacks. The *Attack-Domain* model captures the CDA and supporting (infrastructure) elements that could serve as the target of an attacker. The *Attacker* model captures adversary's intentions as functions on the attack-domain model. This helps identify the target and intent of an attacker. The *attack* model captures the relevant encompassing elements of an attack which include: Attack Domain, Attack Vectors, Attack procedures and the start/end states of the attack. The *ACA* Framework summarised in Fig. 6.1 is a process consisting of eight stages, labelled 1 through 8, in order to derive attacks. Similar to [42] at stage 1 is an abstract[3] domain model. Three dimensions can be used to describe the attack-domain space: components Cm, process properties *Pr*, and system performance metrics Pe. For example, Cm contains terminal nodes, communication nodes, etc.; while *Pr* contain properties such as number of trades, surplus(profit margins) etc.; and Pe contains metrics such as allocative efficiency, number of messages, time to trade, etc.; but no further details of these elements (Fig. 6.2). Stage 2 is a mapping of the abstract attack-domain maps to a concise attack-domain model for a CDA as explained in Sect. 6.2.1. Stage 3, 4, 5 are aimed at defining the attacker's behaviour aspects. Stage 3 defines the *Attacker Intents* which are combined with the concrete domain model in stage 6 to generate an attacker model. Stage 4 and 5 defines the *Attacker's Capability* model and *Attackers Rational* respectively. In stage 7 the *Attacker* model is combined with the states, Attacker Capability, Attack Vectors, and Attack Procedures to generate an *Attack* model. The left-hand side inputs are mostly technical aspects, while the right-hand side inputs are human behaviour aspects. Attacks are derived from the attack model and the *Attackers Rational* in stage 8. The following subsections outline the 8-stage process citing specific examples of a decentralised CDA [5] designed for a resource constrained smart micro-grid context.

6.2.1 Attack Domain Model

The attack domain comprises three finite sets namely the component set (*Cm*), property set (*Pr*), and performance set (*Pe*). As shown in Fig. 6.2, each of these sets can be treated as a dimension in a three-dimensional attack space. Formally, an attack-domain model ADM of a CDA is a 3-tuple of (Cm, Pr, Pe), where *Cm*, *Pr*, and *Pe* denote system components, system properties, and system performance metrics, respectively. Thus, ADM defines a finite attack space an attacker can explore and enables specification of the attacker intent.

We understand that Cm includes elements that may be physical, cyber, and logic, but in this book chapter we narrow *Cm* to define cyber components (eg, wireless network) on top of which a CDA algorithm can be run. A CDA is supported by components that are usually networked together to ensure multiple buyers and sellers participate in the auction. Cm defines the cyber and logical components that supports the auctioning algorithm and the participants such as terminals for participants; networking components; software agents trading on behalf of the participants; etc. There certainly are other categories of components in a CDA; thus the examples given herein are not claimed to be complete in any sense. In

[3]The domain is considered abstract as its elements do not have the specifics required for the modelling and analysis of a CDA.

Adepu et al. [42], each element of Cm is also referred to as an attack point, or simply as a point. An attack point serves to define an attack vector.

The Pr dimension includes measurable properties of the products being produced or controlled by the CDA such as market surplus, number of trades, the market equilibrium price. It is important to note that the same property is likely to be measured at different physical/logical locations depending on the architecture of the CDA.

The Pe dimension refers to one or more performance characteristics of a CDA such allocative efficiency, convergence rate to equilibrium price, communication overheads, time of a single trade, etc. While the Pe metrics may appear similar to Pr properties; it is arguably better to consider them as separate dimensions of a CDAs. Consider an element in Pe to refer to a measurement taken at a specific point in the progression of a CDA to ascertain how a component or the whole system is performing. As such elements of the set Pr show how well the whole CDA or its components are performing. Pe can be measured at different physical/logical locations (measurement points) depending on the architecture of the CDA.

6.2.1.1 Decentralised CDA Algorithm

In order to come up with a *concise Attack-Model* there is need to understand the CDA component in the framework. This subsection outlines the different states and activities of a CDA that guide development of attack procedures. While there exist many variants of grid-based CDAs, to support our thesis in this book chapter we consider a decentralised CDA proposed in [5] that is assumed to run efficiently on RCSMG architecture developed by Kayem et al. [1]. For a detailed background on the fundamental concepts of continuous double auction algorithms and their application in grid-like platforms the reader is referred to the following literature [10, 12, 19, 43–46]. Formally, the descriptor of the decentralised CDA is a septuple:

$$P_{CDA} =< \rho, B, Ş, V_b, C_s, \Delta_{price}, t_{round} >,$$

where:

- ρ is the power in single units to be auctioned;
- $\beta = b_1, \ldots, b_n$ is the finite set of identifiers of buyer *TA*s, where n is the number of buyer *TA*s;
- $Ş = s_1, \ldots, s_m$ is the finite set of identifiers of buyer *TA*s, where m is the number of seller *TA*s;
- $V_b = (V{*}_1, \ldots, V{*}_n)$, where $V *_i (v_{i1}, v_{i2}, \ldots, v_{in_i})$ is a vector of unit valuations of *TA* b_i. Here, n_i is the number of units of p that b_i requires, and v_{ij} is the valuation value for the jth unit acquired;
- $C_s = (C{*}_1, \ldots, C{*}_m)$, where $C *_i (c_{i1}, \ldots, c_{im_i})$ is a vector of unit costs of TA s_i. Here, m_i is the number of units that s_i wants to sell, and c_{ij} is the cost of the jth unit;

- Δ_{price} is the minimum price step required in the auction. That is, a buyer (seller) TA must increase (decrease) its bid (ask) at $nx\Delta_{price}$, where n is a non-negative integer;
- t_{round} is used for defining the condition for terminating the CDA; that is, if there are no new asks or bids during a time period t_{round}, or the maximum threshold of rounds per day R is reached, the CDA terminates.

Algorithm Activities
The main activities (valid states, S, see Sect. 6.2.5) in a decentralised CDA can be summarised as follows:

1. Registration: In order to participate in the auction, agents must first register (with a registration manager). This is a once-off procedure. Once an agent is registered, it is able to participate in any number of auctions rounds.
2. Initialisation: At the beginning of each R, *TA*s enter a "mock marketplace" to determine the equilibrium price and overall surplus distribution. The token (a mobile object) is initialised, such that $r = 0$. After this initialisation, when the equilibrium price is found, all trades will commence at this price.
3. Participation Request: A new round of the CDA starts, $r = r + 1$, $oa = \infty$, and $ob = 0$. Any agent that wants to submit an offer in the market will request for the token. A MUTEX protocol is used to serialise a fair market access to the participants.
4. Bid Formation: An agent that receives the token will compute their offer (bid/ask) using the AA trading strategy and submit it into the token.
5. Transacting: Considering the offer is not invalid, if a matching bid/ask is found the transaction is instantly concluded and a trade occurs. Otherwise, the offer is put in the order-book as an outstanding ask/bid. Thus, several situations might arise during a round:

   (a) When a seller *TA* submits an $aska$,

      - if $a \geq oa$ then a is an invalid ask;
      - if $ob < a < oa$, then oa is updated to a;
      - if $a \leq ob$, then this seller *TA* makes a deal at ob; go to 3.

   (b) When a buyer *TA* submits a bid of b,

      - if $b \leq ob$, then b is an invalid bid;
      - $ifob < b < oa$, then ob is updated to b;
      - if $b \geq oa$, then this buyer *TA* makes a deal at oa; go to 3.

   (c) This process repeats until no new bids (asks) are submitted during a time period t_{round}.

   The trade information and outstanding offers are made public and visible to other agents. The number of trades and wins for each *TA* are recorded and kept in the token. We consider 'truthfulness' similar to [43], where both buyers and sellers full-fill their contractual obligations once bids are matched.
6. Termination: The order-book is mutual exclusively distributed until the end of a trade day.

Algorithm Procedures
In [5], the decentralised CDA algorithm has been partitioned into procedures which can be executed at the different components within the auction infrastructure (Figure 6.2). The proposed decentralised CDA protocol results in a mutual exclusion problem. Consider n trading agents (*TA*)s, where ($n > 1$) requests to submit an offer(ask/bid) in the CDA market in order to transact. Since at most, one *TA* can submit an offer in the auction market at a single instance of time, serialization of market access and fair chances to submit an offer was proposed. Two types of request messages are passed by nodes requesting to participate in the auction market: Req_{sm} (global token request) and Req_{mp} (local token request). Incoming Req_{sm} are enqueued in a FIFO $RQ1$ queue while Req_{mp} are enqueued in the $RQ2$ FIFO queues at the M_{sm}. At the M_{sm} nodes is a *POINTER* variable which stores the location of an M_{sm} in possession of the token, or next intermediate M_{sm} pointing to that token holding node (see [47]) where Req_{sm} is sent. When a Req_{sm} is sent, the $TokenAsked$ boolean variable is set to *TRUE* avoiding continuous request messages to be send for the token by the same node. On arrival of the token to M_{sm} and M_{mp} the boolean variables $FlagM_{sm}$ and $FlagM_{mp}$ are set to *TRUE* indicating possession of the token. *GQ* which is a FIFO queue at M_{sm} stores a copy of requests submitted and "locked-in" at the arrival of the token to the cluster head. *TokenOB* is an online copy of the CDA order-book carried in the token. *LocalOB* is a local copy of the CDA order-book updated each time a trading agent participates in the auction market. Each M_{mp} has a *ClusterDir* that contains a directory of neighbouring M_{mp} nodes. A *TokenCounter* keeps record of number of auction market rounds. When the predefined number of rounds is reached trading is terminated and an end-of-trading-day (t_{round}) message may be communicated to the rest of the participating nodes. This message includes trading-day market information.

Local Market Execution Procedure
In this book chapter we focus in the Local Market Execution procedure (Algorithm 6.1) as it is more informative on the types of attacks that can occur. This procedure is executed on receipt of the token at M_{mp}. The $TokenCounter$ variable is incremented and the $FlagM_{mp}$ is set to $TRUE$. The *TA* forms an offer (bid/ask) which is submitted into *TokenOB* and *TokenCounter* is incremented. If the predefined number of rounds is reached, the trading day is terminated. Success or failure of an *TA* s offer to result in a trade does not affect passing-on of the token. The Algorithm 6.1 presents the $LocalMarketExecution$ Procedure.

Sketch Proof: Algorithm 6.1 is a sub-algorithm of a more detailed CDA scheme discussed in depth in [5]. The decentralised CDA therein, is shown to operate and without violating the Mutual Exclusion (MUTEX) protocol it is built upon. To verify correctness of each processing structure of the algorithm, we consider the sketch proof using Hoare logic. Let us denote the problem preconditions by P of the algorithm A, giving the postconditions Q. Say P and Q are the problems precondition and postcondition respectively, we can represent this as follows: If a condition, c is well defined (it can be evaluated), $P \wedge c \xrightarrow{A} Q$ and

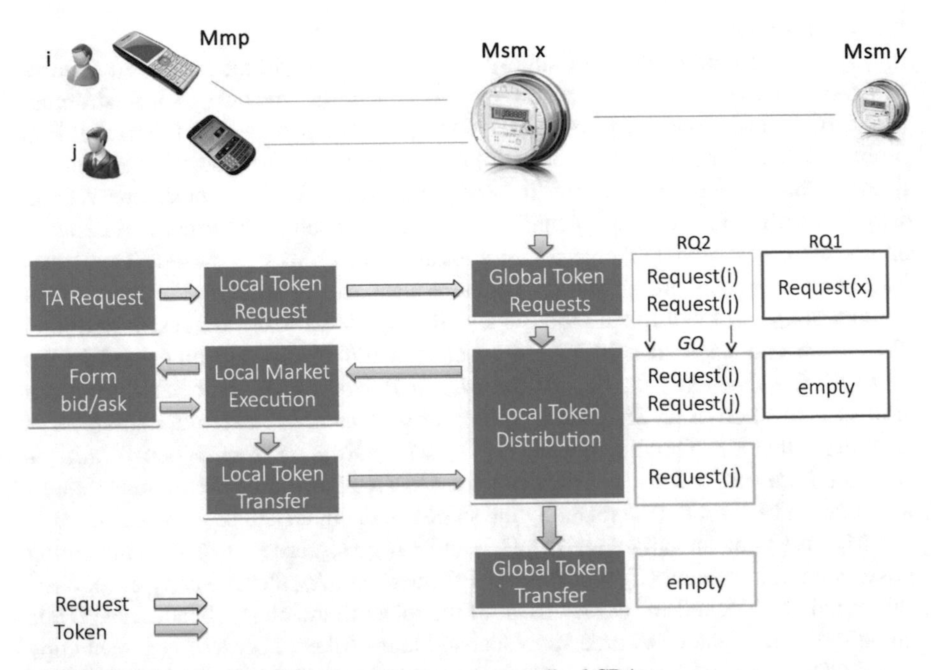

Fig. 6.2 A classification of cheating attacks on decentralised CDAs

$P \wedge \bar{c} \xrightarrow{A} Q$, then $P \xrightarrow{A} Q$. The sub-algorithm's preconditions (considering the main conditional statements line 5) are: $TokenReceived = \{0, 1\}$, $FlagM_{sm} = \{0, 1\}$, $TokenCounter \in \mathbb{R}$, $R \in \mathbb{R}$; where the condition c, $TokenRecieved = TRUE$, $FlagM_{mp} = TRUE$ holds, and the sub-algorithm's postconditions $LocalOB$, $TokenOB$, R and $TokenCounter$ are updated (line 30).

6.2.2 Attacker Behaviour

The ability to model the 'attacker behaviour' can show the attack scenarios that are more likely to happen, which results in more precise risk assessments and damage predictions. Representing attacker behaviour in terms of attack effects instead of the attack itself enables the system security to be indirectly evaluated by identifying families of attacks rather than individual instantiations [48]. Attacker behaviour can be modelled as a strategic decision-making process that accounts for the following factors affecting the attacker's decisions:

Algorithm 6.1: *LocalMarketExecution* Procedure

Input : $TokenRecieved, TokenCounter, R$
Output: $R, TokenCounter\ TokenOB.LocalOB$

```
1  Initialisation: R ← 1000, TokenCounter ← 0
2  for each M_mp ∈ M_mp,a[.] do
3      if TokenRecieved = TRUE AND FlagM_sm = TRUE then
4          TokenCounter ++ ;          // The Token Counter is incremented
5          Set FlagM_mp to TRUE ;
6          Ask TA_i to FormOffer() ;                  // A TA submits its offer
7          if tradeID = buyer then
8              bid ← offer ;
9              if bid ≤ ob OR out of [P_ll, P_ul] range then
10                 bid is invalid ;
11             else
12                 ob ← bid ;
13             if ob ≥ oa then
14                 P_t ← oa ;     // Where P_t is the transaction price at
                   time t
15                 Trade and TokenOB update ;
16         else if tradeID = seller then
17             ask ← offer ;
18             if ask ≥ oa OR out of [P_ll, P_ul] range then
19                 ask is invalid ;
20             else
21                 oa ← ask ;
22             if ob ≥ oa then
23                 P_t ← ob ;     // Where P_t is the transaction price at
                   time t
24                 Trade and TokenOB update ;
25         else
26             no new oa or ob in a pre-specified time period ;
27             Round ended with no transaction ;
28         Update LocalOB from TokenOB ;        // Local Orderbook copy is
           updated
29         R −− ;          // Auto-decrementing remaining trade rounds
30         return TokenOB, R, TokenCounter ;
31     else
32         wait for TOKEN ;
```

6.2.2.1 Attacker Intent

Intent-based approach is evident in earlier design attacks [49]. As social criminals, in cyberspace, attacks are typically not random, and an attacker launches an attack to achieve some malicious goals [50]. An attacker can design an attack with the possible intention of damaging a specific component in its domain without, in the short term, affecting any system property or performance. Alternatively, the attacker may attack by changing some system property such as profit margins (in the case of cheating). Further, the attacker may craft an attack aimed at reducing some system output such as convergence to an equilibrium price within an auction trade day. Such an attacker model captures the mapping of many, perhaps not all, attacker intentions to the attack-domain.

6.2.2.2 Attacker Capability Model

Formally, the attacker capability model *ATC* of a CDA is a 3-tuple (Sk, De, Dn), where Sk, De, and Dn denotes sets of system knowledge, disclosure resources, and disruption resources. Thus, *ATC* defines a finite attack space an attacker can explore in relation to attacker capability/ constraints.Similar to [51, 52], we propose three dimensions for the attack space: the adversary's a priori system model knowledge Sk; his disclosure resources De; and disruption resources Dn. An attacker with *a priori* system knowledge (including knowledge of elements on ADM) can construct more complex attacks, possibly harder to detect and with more severe consequences. This depends on the extent and depth of the knowledge such as knowledge of terminals used by auction participants; the networking components; the software components etc. Similarly, the disclosure resources De enable the attacker to obtain sensitive information about the system, or its individual components during the attack by violating the data confidentiality. For example the attacker can obtain private information secret to other auctioneers (trading agents). We consider that disclosure resources cannot be used to disrupt the system operation. On the other hand, disruption resources can be used to affect the system operation, which happens for instance when data integrity or availability properties are violated.

6.2.2.3 Attacker Rational

As emphasised by Sedaghatbaf and Abdollahi Azgomi [53], it is necessary to find out how probable that attack is, from its potential costs and benefits, instead of relying only on the technical aspects of the system. The supporting notion is that attacks are considered unlikely if their cost is not worth their benefits. Thus, if an attack is unlikely to occur, its assumed the system is secure towards that attack in spite of its potential technical vulnerability, because it is irrational for an attacker to perform an unprofitable attack. Formally, the Attacker's Rational R is a 3-tuple (Cb, Sd, Ac), where Cb is a cost-benefit-analysis function, Sd is the behaviour set dependent on the system's defence, and Ac is the number of attackers and the coordination between them. This model is helpful in building a taxonomy of attacks within the *ACA* Framework.

- *Cost-Benefit-Analysis:* An attacker's decision to execute an attack scenario is made on the basis of some cost-benefit model such as those defined in [54, 55]. Thus, we assume that an attacker attacks only if the overall attack is profitable, and if so, he chooses the most profitable strategy in each stage of the attack. This assumption serves as an upper bound in the modelling context, as any irrational decision hinders the end result of the attacker himself.

- *System Defence Interdependency:* The behaviour of an attacker and the system's defence mechanism are interdependent, and security reactions can change the possible attack steps for the attacker [50]. Petri net-based models [56] or partially observable Markov decision process [48] can be used to capture the dynamics between the interaction between attacker and defender (i.e. the system's defence mechanism).
- *Attacker Coordination:* The type of plausible attack need to factor in the potential of more than one individual attacking the system concurrently. Such attackers can be individualistic or can collaborate and coordinate to deliver some attack that benefits the parties.

6.2.3 Attacker Model

The *Attacker Model* Ä is defined as a 2-tuple (Ĩ, ADM) where Ĩ is a finite set of attackers' intents and ADM is an attack-domain model. In this chapter, the attacker's goal or an objective is treated as an intent. Given its inherently personal and social nature, Adepu and Mathur [42], note it is difficult to precisely specify the intent set Ĩ . However, for this book chapter we consider intents to include: damage to a component (Cm), disturbing a system property (Pr), or altering system performance (Pe). An intent is applied to an element or a region of the domain. Thus, an intent can be considered as a function that is applied by an attacker to one or more elements of an *Attack Domain* that defines a CDA. The *Attacker Model* does not describe the actual attack itself. Attacks have to be designed to achieve a goal inherent in an intention.

6.2.4 Attack Model

For a CDA, say X we consider an attacker model $\ddot{A}_X = (\tilde{I}, ADM)$, where $ADM = (Cm, Pr, Pe)$ and Ĩ is a function. The attack model is a sextuple where, Å is potentially infinite set of procedures to launch attacks; Č is a attacker capability model for the attacks derived from *ATC*; Ĩ is a finite set of attacker intents; D is the domain model for the attacks derived from the attack-domain model *ADM* of X; $P \subseteq Cm$ is a finite set of attack points; and S_0 and S_e are possibly infinite sets of states of X, that denote, possible start and end states of interest to the attacker; R is the attacker rational that determines the feasibility of an launching the attack. An attack point in X could be a physical element or an entry point through the communications network connecting CDA and the underlying power system.

We are aware that some attacks could be modelled to leverage on faults, but that phenomenon is beyond the scope of attacks we cover in the *ACA* Framework. Implicitly, our attack model herein assumes no faults occur (thus, fault-related attacks are excluded). We note that several formalisms exist for modelling attack procedures, Å, by employing graphical methods, among others [56, 57], and [58] as cited in [42]. Modelling of Å is not addressed in this chapter. Å cannot be enumerated unless constraints are imposed on the attack methods. The set Å is similar to the attacker payload [49]. From the above *Attack Model* definition several attack models can be built for a given CDA. The different attack models on the same CDA are derived by altering/changing the constraints imposed on Å and selecting different subsets of the Attack-Domain variables (Cm, Pr, Pe); and Attacker Behaviour which encompasses the Attacker Capabilities (Sk, De, Dn); the Attacker Rational (Cb, Sd, Ac). The context in which an attack model is used will determine its appropriateness and effectiveness. We expect the attack model presented herein, to be useful in designing, among other attacks, cheating attacks to study the resilience of a decentralised CDA.

6.2.5 Attacks

For a CDA, say X we consider an attack model

$$Æ_X = < Å, \check{C}, \tilde{I}, D, P, S_0, S_e, R >$$

An attack Æ in X is defined as a terminating or a non-terminating procedure å ∈ Å designed to realize a finite set of intents ĩ ⊂ Ĩ , aimed at domain $d \subset D$, requiring a finite set of capabilities č ⊂ Č, launched through a finite set of points $p \subset P$ when X is in state $s_0 \in S_0$ and possibly removed when X is in state $s_e \in S_e$.

Attack Success
We consider an attack Æ to be successful if all intents in ĩ are realized in a finite time. Part successful attacks occur when a subset of intents in ĩ are realized, while unsuccessful attacks occur when none of the intents in ĩ is realized and there is no intended or unintended side effect of applying an å to CDA_X. The attack procedure å may or may not terminate after all intents in ĩ are realized.

Attack Vector
An attack vector is a path in a CDA that starts at an attack point p and allows the exploitation of a vulnerability. In other words, an attack is a parameterised procedure that exploits such a path. Identification of attack vectors is beyond the scope of this book chapter, but it is possible to identify attack vectors for the design of attack procedures using the attack domain model and knowledge of its operation.

Attack Procedure
One key factor to a successful attack is the design of an effective attack procedure å. This requires the attacker to be familiar with at least the targeted components and CDA system C. In cases where multiple points are attacked and deception is needed

to avoid detection, å might actually involve computation of data values being sent in real time. For example an attacker may be interested in knowing the evaluation and reservation prices of other victim participants in real time, so as to inform his decision in the CDA. Doing so requires a deeper knowledge of the properties of the CDA and its components. It is assumed that such knowledge is used in the design of å for a successful attack. Thus, in the attacker and attack models, the knowledge of the system: partial or complete (as specified in Attack Capability model), is encapsulated in the attack procedure.

Attack Start States
Say an attacker is able to install a piece of malware into a component(s) when the CDA system is in some state s. Then consider the malware will remain dormant until the CDA system reaches another state $s^{'}$ when it actually executes the intended attack. The start state s_0 in this case can one of who cases:

- The attack is a sequence of two separate attacks where malware injection attack followed by the malware's payload execution. The start state for malware injection is s and for the subsequent payload execution launched by the malware is $s^{'}$.
- The entire attack is one and we treat the start state as s.

The best approach between the aforementioned two depends on the intent of the attack. If the state in which malware is inserted is key to the attack being not detected, while the payload execution can be launched in any state, then s should to be considered the start state while $s^{'}$ can be ignored. However, if malware must launch the subsequent attack in a specific state for intent realization, then $s^{'}$ should be considered. Examples of attacks and their types are given in Sect. 6.3.

Valid State Sequences
Valid sequences are those sequences that appear at least once during the normal operation of CDA_C (see Sect. 6.2). Given how complex a CDA can be, the space of all valid sequences of any arbitrary length is huge and difficult, if not impossible, to enumerate. Under attack, or in case of some form of failure, CDA_X might enter an invalid sequence, i.e., a sequence that would never occur under normal operation constrained by rules of the CDA.

6.3 Cheating Attack Cases

In this section, we use the *ACA* Framework as a basis for modelling the cheating attacks towards a decentralised CDA. The cheating attacks we consider herein are not exhaustive but crucial in informing on the mitigation measures that can be put in place ensuring the CDA algorithm executes successfully. The *Victim Strategy Downgrade* and *Collusion by Dynamic Strategy Change* have been described in [7]. We expand on the aforementioned attacks with a more formalised approach demonstrating how the proposed framework can be used in designing such attacks.

Additionally, we consider *Profiling and Market Prediction Attack*; and *AA Manipulation Attack*. The aforementioned attacks are chosen because they:

- can be generalised and can provide a wider coverage of attacks to consider for the system defenders;
- can provide more insight into how the system will behave under different attack procedures and parameters;
- can be used to learn and test the resilience of the system.

6.3.1 Victim Strategy Downgrade

An attacker, P_{Ai}, where ($i = 1$, $P_A \subset P$, and P is a set of all participants) employs an automated tool (e.g. malware) to gain control of other trading agents *TA* strategies. The attacker can 'downgrade' victim *TA*s to trading strategy that is inferior enabling them to obtain additional surplus. We are aware the same attack can be perpetrated by an adversary simply changing their agent strategy to adopt a different but superior bidding strategy which affords higher surplus. We will focus on the former, as it presents an exciting problem opposed to a simple P_{Ai} upgrading to a superior strategy. The Victim Strategy Downgrade has been described in detail in the literature [7] and [8]. The term attacker and adversary are used interchangeably. We shall consider the following assumption:

- the adversary P_A has control of $P - 1$ victim *TA*s;
- despite $P - 1$ victim *TA*s having their strategy downgraded, to say, *ZI Strategy*, the *TA*s will seamlessly continue participating in the auction as normal;
- the adversary has computational capacity to launch the attack and a reliable connection exists with the victim *TA*s

Motivation
A number of trading agent strategies that have been developed for the CDA over the years. We note experimental evidence that supports the phenomenon of some strategies being superior; with the ability to gain more surplus from trade than their inferior counterparts [18, 19, 59, 60]. The Adaptive Aggressive strategy is one such superior strategy, while the Zero Intelligence is the most inferior [19, 59, 60]. Ma and Leung [18], demonstrates that AA agents are adaptive to different combinations of competitors; and to different supply and demand relationships. ZI agents behaved worse since they do not analyse their environment and the other agents whom they are competing with. The AA obtains huge profit margins in comparison to ZI strategy [18].

Attack
Assume an attacker distributes a malicious code to attach itself onto other participants *TA*s (the attack vector). The malicious code carries a payload capable of incorporating an inferior *ZI* strategy to the victim *TA*s' bid forming mechanism.

Formally, a strategy downgrade attack Æ(sda) in X is defined as a terminating procedure å ∈ Å designed to realize a finite set of intents ĩ ⊂ Ĩ (where ĩ includes gaining control of other *TA*s strategy component, changing their strategy, influencing the market to gain additional surplus), aimed at domain $d \subset D$ (where d is a set of all $M_m p$ components where *TA*s are hosted), requiring a finite set of capabilities č ⊂ Č (č includes knowledge of participants, ability to launch a malicious code,[4] access and communication with victim *TA*s), launched through a finite set of attack points $p \subset P$ (p is a set of all victim *TA*s), when X is in state $s_0 \in S_0$ (where s_0 is a set of all valid CDA states where an attack can be launched) and possibly removed when X is in state $s_e \in S_e$ (where s_e is a set of all valid states the attack can be stopped which is controlled by a clock trigger).

Attack States
The attack is a sequence of two separate phases: malware injection phase followed by the malware's payload execution phase. The start state for malware injection is s and for the subsequent payload execution launched by the malware is s'. Malware injection occurs in any of the valid states of the CDA, while payload execution is triggered by a clock. It is important that victim *TA*s are all in s' concurrently through a synchronisation protocol. The aim of this attack is not to put the overall CDA into an invalid state, but to ensure despite the attack the CDA is always in its valid states.

Attack Variants
This attack can take two forms: static or dynamic downgrade. In static downgrade, an agent will instantly change its strategy on infection (once adversary payload is delivered). This change is somehow permanent. If no further coordination occurs between the infected agents and the adversary, it would be difficult to use communication overheads to infer occurrence of such cheating. However, such attack can easily be detected by analysis of market efficiency as *ZI* agent population yields fairly lower market efficiency than a homogeneous *AA* population. Thus, an advanced attacker would employ a dynamic downgrade to victims *TA* strategies allowing victim agents to revert to the *AA* strategy based on a clock-based trigger. We consider the payload incorporates the inferior *ZI* strategy to the victim *TA*s bid forming mechanism. Since the attack is dynamic, the payload will ensure the victim toggles between *AA* and *ZI* strategy in strict response to a clock trigger or messages from the adversary. Algorithms 6.2 and 6.3 present the dynamic attack from an attacker and the victims' perspective.

Attacker Rational
In order to evade easy detection an attacker can send $P - 1$ 'revert' messages to the victims using a private back-channel directly to victims. Only the trading outcome is affected as the CDA remains fairly the same. The Adversary is required

[4]These are described as disclosure resources.

to be computationally apt, and to establish a reliable connection with the victims in order to execute such an attack. In a resource constrained setup, this might not be feasible. To maximise on the surplus to be gained the adversary can only participate when the clock trigger occurs. If a considerable number of victim *TA*s have requested for the token, the adversary might not gain as high a surplus as early requesting *TA*s will have traded against each other. This can be addressed by making the clock trigger to encompass the full duration of a single or multiple trade days randomly.

Attack Analyses

The adversary P_A is required to be computationally apt, and to establish a reliable connection with the victims in order to execute such an attack. In a resource constrained setup, this might not be feasible. To maximise on the surplus to be gained the adversary can only participate when the clock trigger occurs. Arguably, if a considerable number of victim *TA*s have requested for the token, the adversary will not gain as high a surplus as early requesting *TA*s will have traded against each other. This can be addressed by making the clock trigger to en-campus the full duration of a single or multiple trade days randomly.

Algorithm 6.2: Dynamic Strategy Downgrade Attack—Adversary

Input : $WillingToTrade, Clock$
Output: $Revert_{msg}$, $ReqM_{mp}$

```
1  Initialisation: R = 1000, Clock = FALSE
2  /* Adversary sends a payload triggered by a clock                 */
3  repeat
4      repeat
5          Wait;
6      until Clock = TRUE;
7      if WillingtoTrade = TRUE then
8          Request TOKEN ;
9      else
10         Send Revert_msg to P − 1 victims ;
11 until termination;
```

Sketch Proof: Algorithm 6.2 outputs $Revert_{msg}$ messages to the victim *TA*s if the adversary is not willing to trade and the clock triggers a strategy downgrade. The sub-algorithm terminates at the end of a trade day when $R = t_{round}$. The loop will always terminate when $Clock$ is TRUE (since the attack is clock triggered). If we consider the inner conditional statement (line 7), say S **if c then** A_1 **else** A_2 **endif**. If c is well defined (meaning it can be evaluated), $P \wedge c \stackrel{A}{\rightarrow} Q$ and $P \wedge \bar{c} \stackrel{A}{\rightarrow} Q$, then $P \stackrel{A}{\rightarrow} Q$. This suggests that we verify the correctness of the branch (both when c is true and when c is false). The sub-algorithm's preconditions are $WillingTotrade = \{0, 1\}$, $clock = \{0, 1\}$ and the postcondition is a $Revert_{msg}$ or $ReqM_{mp}$. If the condition c is $WillingToTrade = TRUE$ then a request is

Algorithm 6.3: Dynamic Strategy Downgrade Attack—Victim

```
   Input : Revert_msg, Clock
   Output: strategy
1  Initialise: R = 1000, Clock = FALSE
2  /* For each infected TA                                   */
3  for each TA ∈ TA_i[.] do
4  |   repeat
5  |   |   repeat
6  |   |   |   Wait;
7  |   |   until Clock = TRUE;
8  |   |   repeat
9  |   |   |   Use ZI strategy;
10 |   |   until Revert_msg = TRUE or Clock = FALSE;
11 |   until termination;
```

made for the TOKEN. In the other case, $WillingToTrade = FALSE$ holds and a $Revert_{msg}$ is sent to the victim *TA*s. In Algorithm 6.3 each victim will change strategy when the clock trigger is TRUE. This repeats until $Revert_{msg}$ is received or the clock trigger is off (FALSE).

6.3.2 Dynamic Collusion Attack

Assume adversary participants P_{Ai}, form a coalition and coordinate among themselves using an automated tool to gain additional surplus over the rest of victim participants, P_{ν}. We consider, $i = \{1 \ldots \eta \ldots \kappa\}$, where η is the bound on maximum colluders that guarantee added profit on strategy switch, κ is the number of all the colluders and $P_A \subset P$ and $P_{\nu} \subset P$. The aim of the tool is to coordinate population ratio of P_A's to P_{ν}'s by maintaining η. The P_{ν} will continue using the default strategy (in this case the AA strategy). For this *Collusion Attack* we further consider the following:

- a symmetrical relationship of buyers and sellers;
- all *TA*s get the same amount of units to trade;
- only the P_A colluders can change their strategy;
- despite random and numerous strategy changes, P_As can seamlessly continue participating in the auction;
- the P_A colluders have computational capacity and reliable channels to coordinate the attack.

Motivation
Vach and Maršál experimental results in [61] show that additional surplus can be gained from changing population ratios of *TA*'s from *AA* to *GDX*[5] strategy with the minority population dominating in average profit.

Attack
κ of P_A install a tool (piece of software or script) that enables only η to autonomously change agent strategy from *AA* to *GDX*. Colluders' TA is the attack points where the CDA rule allowing TAs to participate regardless of the bidding strategy they use becomes the vector of this type of attack. The colluding *TA*s use a separate channel to communicate among themselves. On receipt of the beginning of trading day signal the strategy changing automated tool will allow a number of adversary agents to shift their strategy to the *GDX*. In considering experimental evidence in literature [61], the adversary agent population ratio to the truthful agent population can be either 2 to 4 or 1 to m to ensure a high surplus on adversary population. To ensure such coordination the strategy changing automated tool can use some group MUTEX protocol to select η number of colluders allowed to change their strategy at one particular time. The η-MUTEX algorithm will allow at most η colluders at a time to change their strategy (enter critical section). The protocol is token based and η tokens are used. Thus, a colluding *TA* can only change its strategy to GDX when it is in possession of the token. As an input the algorithm gets P; the number of all participants in the market. For instance, Chaudhuri and Edward [62] proposed one such protocol which performed on the worst case message complexity of $O(\sqrt{n})$. The single trade-day-signal that is used to trigger the selection of colluding *TA*s can be altered to a number of signals (therefore trading days) to allow the selected colluders more number of rounds to benefit before the change. Tolerance to node and link failure ensures robustness of the η-MUTEX protocol. We assert that as long as the colluders are coordinated in such a manner this attack will not deviate, with high probability; κ colluders will continuously take turns and cheat.

Formally, a dynamic collusion attack Æ(dca) in X is defined as a non-terminating procedure å ∈ Å designed to realize a finite set of intents ĩ ⊂ Ĩ (where ĩ includes coordination with other colluders *TA*s strategy component, using a κ-MUTEX to switch between strategies), aimed at domain $d \subset D$ (where d is a set of all $M_m p$ components where colluding *TA*s are hosted), requiring a finite set of capabilities č ⊂ Č (č includes knowledge of all participating *TA*s, ability to launch a coordinating tool, connection and communication with colluding *TA*s), launched through a finite set of enabling points $p \subset P$ (p is a set of all colluding *TA*s), when X is in state $s_0 \in S_0$ (where s_0 is the initial state CDA at the start of an auction where an attack is launched) and possibly removed when X is in state $s_e \in S_e$ (where s_e is a set of all valid states the attack can be stopped or other colluders are selected).

[5] Strategy developed by Tesauro and Bredin [29] as a modification of the Gjerstad-Dickhaut (GD) strategy that uses dynamic programming to price orders.

Attacker Rational
Intuitively, a group of attackers can opt collude motivated by the significant surplus there will obtain in the CDA. In taking turns to cheat, this attack can be difficult for the system defenders to detect. Since at any time, κ colluders will cheat and obtain surplus, the system defender can be suspicious of such behaviour. A possible workaround would be to randomly allow such cheating in different periods of the auction market.

Attack Analysis
Additional computational overheads are incurred by the κ adversaries, as they need to establish a reliable connection among themselves and change their strategy in turns. No additional messages are introduced on the CDA communication channel in this attack, assuming that adversaries strictly communicate in a back channel. Using the same experimental findings that can be used to support occurrence of this collusion attack [61], we observe that if all users do not use *AA* strategy the market allocative efficiency is significantly reduced. Additionally, messages need to be exchanged among the κ P_As. For this attack to be a success, at least η adversaries ($\kappa = \eta$) should agree to collude, otherwise additional surplus will not be realised. This is because the ratios between *AA* (of victim *TA*s) and *GDX* (of adversary *TA*s) is such that *GDX* can not gain additional surplus.

6.3.3 Evasive Agent Attack

We consider a single adversary that employs an evasive strategy that leverages on other traders secret information such as reservation price to make better predictions of bids/asks to submit. Assuming the auction allows trade of single indivisible electrical power/energy unit. Each bidder and seller associates two values with a unit of energy—a reservation price and a bid/ask. Reservation price (limit price) is the maximum (minimum) price a bidder (seller) is willing to pay (be paid) for energy based on personal valuation and preferences. This information is private to each trading agent. An offer (bid/ask) on the other hand is the publicly declared price that a bidder (seller) is willing to pay (sell). We further consider the following assumptions:

- a symmetrical relationship of buyers and sellers;
- all *TA*s get the same amount of units to trade;
- only the P_A can change its strategy;
- despite random and/ or numerous strategy changes, the attacker P_A can seamlessly continue participating in the auction;
- the attacker P_A and the victims P_V have computational capacity and a separate reliable channel to send their reservation price to the attacker P_A.

Motivation
Assume each agent's reservation value can be independently drawn from a cumulative distribution function (CDF) F over $[0, 1]$, where $F(0) = 0$ and $F(1) = 1$. We assume $F(.)$ is strictly increasing and differentiable in the interval $[0, 1]$. The derivative of CDF, $f(\lambda)$ is then the probability density function (PDF). The adversarial node knows his reservation value and the distribution F of other agents. The adversarial agent tries to maximize his utility and quits the auction if the auction price goes beyond its reservation value. The private data is then used in formulating a trading pattern to obtain a trading advantage over other traders. This phenomenon is very similar to insider trading [63]. Kyle uses a dynamic model of insider trading to examine the value of private information to an insider [63].

Attack
Similar to the aforementioned attacks, an adversary employs a tool (e.g. a piece of malware) whose payload elicits victim *TA*s' private data (e.g. reservation price) and sends it to the attacker. The vector of the attack is other participants *TA*s. At the beginning of each *trading day* or a predetermined time specified by a timer, the victim *TA*s will automatically send their private information to the adversary agent. We assume the adversarial node employs fuzzy logic where the illegally elicited private information forms part of a fuzzy set to inform some fuzzy rules. Knowledge gained from the fuzzy sets can be combined using rules to make decisions based on this information. One such strategy is proposed by He et al. [30]. This approach is supported by the notion that a *TA*'s decision-making about bidding involves uncertainty, multiple factors, and non-determinism that are affected by the attitudes toward risk of its opponents, the nature of the market supply (demand), and the preferences of the other bidders.

Formally, an evasive attack Æ(ea) in X is defined as a non-terminating procedure å ∈ Å designed to realize a finite set of intents ĩ ⊂ Ĩ (where ĩ includes gaining control of other *TA*s strategy component, eliciting their reservation price, using the data to make informed offers in the market), aimed at domain $d \subset D$ (where d is a set of all $M_m p$ components where *TA*s are hosted), requiring a finite set of capabilities č ⊂ Č (č includes knowledge of all participating *TA*s, ability to launch a malicious code, connection and communication with victim *TA*s), launched through a finite set of attack points $p \subset P$ (p is a set of all victim *TA*s), when X is in state $s_0 \in S_0$ (where s_0 is a set of all valid CDA states where an attack can be launched) and possibly removed when X is in state $s_e \in S_e$ (where s_e is a set of all valid states the attack can be stopped).

Attack States
The attack is a sequence of two separate phases: malware injection phase followed by the malware's payload execution phase. The start state for malware injection is s and for the subsequent payload execution launched by the malware is $s^{'}$. Malware injection occurs in any of the valid states of the CDA, while payload execution is triggered by a clock. It is important that victim *TA*s are all in $s^{'}$ concurrently through a synchronisation protocol. The evasive attack does not put the system in an invalid state, which arguably makes it difficult to detect through simple defensive

measures that observe occurrence of invalid states. Algorithms in 6.4 and 6.5 present the Evasive Attack from the victims' and attacker's perspective.

Attack Variants
In our description of the evasive attack, we only considered a single adversary. however, a variant of such an attack can have multiple non-cooperating attackers using the same attack simultaneously. Intuitively, the steps of the attack are the same but additional aspects should be considered. For instance, the re-infection of victim TAs by other attackers; attackers being victims to attacks from other attackers; the increase in computational and communication overheads from the victims perspective; increase in complexity of concurrent attacks.

Attack Analysis
This attack provokes frequent messages of the victim *TA*'s reservation price to be sent to the attacker. Additional computational resources (memory and processing power) are consumed within each victim node, while additional storage and computational resources are expected at the adversary node. Assuming the victim *TA*s communicate with the adversary through a back-channel(to avoid easy detection), the initial message complexity of the CDA algorithm is not affected. The attack results in frequent messages being passed between the attacker and the victim *TA*s. Such activity can be observed and used to inform some irregular agent behaviour.

Algorithm 6.4: Evasive Attack—Adversary

Input : $WillingToTrade, Clock$
Output: $R, Conceal_{msg}, ReqM_{mp}$

```
1  Initialisation: R = 1000, Clock = FALSE
2  /* Adversary sends a payload triggered by a clock                    */
3  repeat
4      repeat
5          Wait;
6      until Clock = TRUE;
7      if WillingtoTrade = TRUE then
8          Change to FL Strategy;
9          Request TOKEN ;
10     else
11         Send Conceal_msg to P − 1 victims ;
12 until termination;
```

Sketch Proof: The Algorithm 6.4 outputs $Conceal_{msg}$ messages to the victim *TA*s if the adversary is not willing to trade while the clock has triggered a strategy downgrade. The outer loop (line 3), guarantees the sub-algorithm terminates at the end of a trade day when $R = t_{round}$. The inner loop (line 4) will always terminate when $Clock$ is TRUE (since the attack is clock triggered). If we consider the inner conditional statement (line 7), say S **if c then** A_1 **else** A_2 **endif**. If c

Algorithm 6.5: Evasive Attack—Victim

Input : $ClockInput, Conceal_{msg}$
Output: $Private_{msg}$

```
1  Initialise: R = 1000, Clock = FALSE
2  /* For each infected TA                                   */
3  for each TA ∈ TA_i[.] do
4  |   repeat
5  |   |   repeat
6  |   |   |   Wait;
7  |   |   until Clock = TRUE;
8  |   |   repeat
9  |   |   |   Send Private_msg;
10 |   |   until Conceal_msg = TRUE or Clock = FALSE;
11 |   until termination;
```

is well defined (meaning it can be evaluated), $P \wedge c \overset{A}{\rightarrow} Q$ and $P \wedge \bar{c} \overset{A}{\rightarrow} Q$, then $P \overset{A}{\rightarrow} Q$. We verify the correctness of the branches (both when c is true and when c is false). The sub-algorithm's preconditions are $WillingTotrade = \{0, 1\}$, $Clock = \{0, 1\}$ and the postcondition is a $Revert_{msg}$ or $ReqM_{mp}$. If the condition c is $WillingToTrade = TRUE$ then a request is made for the TOKEN. In the other case, $WillingToTrade = FALSE$ holds and a $Conceal_{msg}$ is sent to the victim *TA*s. In Algorithm 6.5 each victim will send $Private_{msg}$ when the clock trigger is TRUE. The outer loop (line 4), shows the sub-algorithm terminates at the end of a trade day, t_{round}. The first inner loop (line 5) will always terminate when $Clock$ is TRUE. This warrants victim TAs to sent until Clock is FALSE or a $Conceal_{msg}$ from the adversary is recorded.

6.3.4 Adaptive Aggressive Strategy Manipulation

Assuming an attacker has access and control of victim TA's through an automated tool (a malware) he/she can manipulate a number of components, parameters and variables involved in the bid-formulation process of a population of homogeneous victim *TA*s. The end-goal in this scenario is to influence victim *TA*s behaviour to get favourable offers in the auction. For instance, victim *TA*s will be manipulated to sell (buy) energy at very low (high) offers favourable to a buying (selling) attacker. Figure 6.3 shows the components (presented in [19]) that an adversary could manipulate in order to tip the auction balance to their favour. Overall, the AA Strategy manipulation attack can take one of many forms, which include: market information attack, adaptive parameters attack, agent preferences attack. These variations are not exhaustive, but, can be helpful in understanding *AA Strategy*

Manipulation Attack. For this chapter we shall consider only two randomly chosen variants of the *AA Manipulation Attack*, namely: *Market information Attack* and *Adaptive-Parameters Attack*.

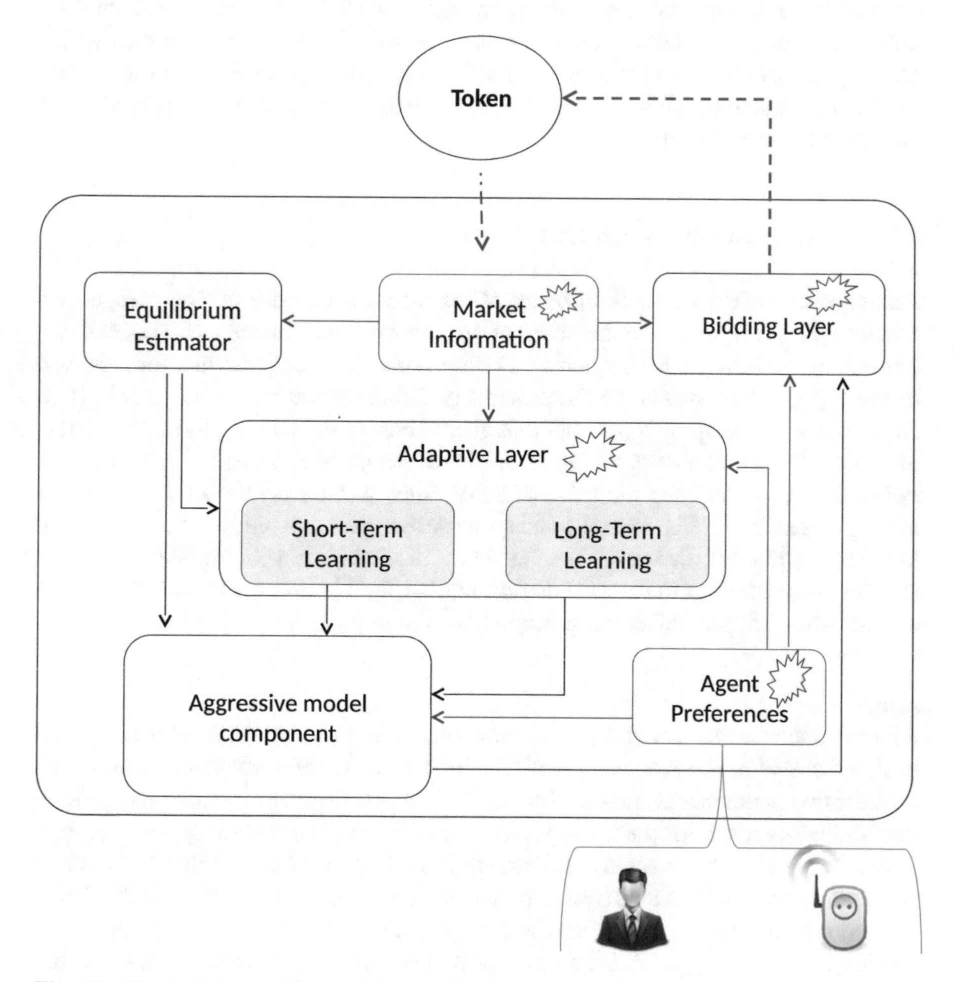

Fig. 6.3 The AA Strategy Manipulation attack points

6.3.4.1 A Market-Information Attack

An adversarial payload can alter the victim TA's market information input to trigger aggressive agent behaviour in the auction. Figure 6.3 shows how the AA strategy's three main components of 'logic and reasoning'(the Equilibrium Estimator, the Bidding Layer and the Adaptive Layer) rely on market information (the current outstanding bid/ask, the equilibrium price, etc.) as input to their computations. The payload will give misrepresented market information as input to the victim *TA*s causing the victim *TA*s to make false (non-profitable or non-competitive) offers. For instance, assume the payload's input of the market conditions is a false high outstanding bid to a buyer *TA*. This means the buyer *TA*s are forced to adapt their bids to the false outstanding bid. The result is a set of non-competitive bids which can be profitable for the adversary. This ultimately prolongs converge of the auction market to a stable equilibrium price. Further, market efficiency is expected to be greatly reduced as a result.

6.3.4.2 Adaptive-Parameters Attack

An aggressive agent will submit better offers than what it believes the competitive equilibrium price to be in an attempt to improve its chances of successfully transacting. In turn, such an agent will compromise its profit margins for a chance to trade [19]. Thus, similar to the previously discussed variant of the attack, if an adversary was to employ a payload that alters short term and long term parameters of victim *TA*s, the adversary will gain an added advantage and surplus in the market. For instance, by changing the value of θ (volatility parameter) the adversary forces trading agents under his control to adapt a more aggressive bidding behaviour,[6] in long term, while his own agent obtains better profitable deal. Intuitively, by failing to effectively adapt to the ever-changing market the victim agents will obtain less surplus while adversarial agent(s) benefit from this condition.

Attack

Assume a single attacker delivers a malicious code to victim *TA*s whose payload is aimed at one of the AA manipulation attack variants (Market information attack or Adaptive parameters attack). The payload can incorporate a malicious code to the victim *TA*s that corrupts the input required for normal AA strategy functioning. A decentralised CDA ensures that market access is granted on the basis of a token which is distributed by a MUTEX protocol. Arrival of the Token (which carries the order-book and ensures a single TA mutually exclusively participates in the auction), can be a trigger that the payload will use to effect changes on the specific AA strategy component. An AA strategy attack, Æ(aasma) in X is defined as a non-terminating procedure å $\in$ Å designed to realize a finite set of intents ĩ $\subset$ Ĩ (where ĩ includes gaining control of other *TA*s strategy component, altering inputs of internal components of TAs, provoke bad offer creation), aimed at domain $d \subset D$ (where d is a set of all $M_m p$ components where *TA*s are hosted), requiring a finite set of capabilities č $\subset$ Č (č includes knowledge of all participating *TA*s, ability to launch a malicious code, connection and communication with victim *TA*s), launched through a finite set of attack points $p \subset P$ (p is a set of all victim *TA*s AA strategy

[6] Aggressive behaving agents focus on successfully bidding while trading off their profitability.

component), when X is in state $s_0 \in S_0$ (where s_0 is a set of all valid CDA states where an attack can be launched) and possibly removed when X is in state $s_e \in S_e$ (where s_e is a set of all valid states the attack can be stopped).

Attack States
The start state for malware injection is s and for the subsequent payload execution launched by the malware is s'. Malware injection occurs in any of the valid states of the CDA, while payload execution is triggered by the arrival of a token at the victim TA. The AA strategy manipulation attack does not put the CDA in an invalid state, arguably making it challenging for the system defender to detect through the occurrence of invalid states. The attack only affects the trade outcome as the CDA states remain the same. Algorithms 6.2 and 6.3 present the dynamic attack from an attacker and the victims' perspective.

Attacker Rational
Similar to the strategy downgrade attack, an attacker can send $P - 1$ 'revert' messages to the victims using a private back-channel directly to victims in order to evade easy detection. For this to occur, the Adversary is required to be computationally capable of establishing a reliable connection with the victims and coordinating the poisoning of their strategy component. In a resource constrained setup, this can be challenging. This attack is similar to the strategy downgrade attack in that it ensures victim *TA*s perform inefficiently in the market, while the attacker with a better efficient strategy capitalises on this phenomenon.

Algorithm 6.6: AA Strategy Manipulation (Adversary)

Input : $TokenReceived$
Output: $Relieve_{msg}$, $ReqM_{mp}$

```
1  Initialise: R ← 1000, TokenReceived ← FALSE
2  /* Adversary sends a payload triggered by a arrival of the
      TOKEN                                                   */
3  repeat
4      repeat
5          Wait;
6      until TokenReceived = TRUE;
7      if WillingtoTrade = TRUE then
8          Request TOKEN ;
9      else
10         Send Relieve_msg to P − 1 victims ;
11 until termination;
```

Sketch Proof: Algorithms 6.6 and 6.7 share similar construct and logic with Algorithms 6.2 and 6.3 respectively. For instance, Algorithm 6.6 has the $Relieve_{msg}$ send to P-1 victim *TA*s when the adversary is not willing to trade, as opposed to the

Algorithm 6.7: AA Strategy Manipulation (Victim)

```
   Input  : TokenReceived, Relieve_msg
   Output: offer
1  Initialise: R = 1000, TokenReceived = FALSE
2  /* For each infected TA                                              */
3  for each TA ∈ TA_i[.] do
4  |   repeat
5  |   |   repeat
6  |   |   |   Wait;
7  |   |   until Tokenreceived = TRUE;
8  |   |   repeat
9  |   |   |   Manipulate inputs;          // Dependent on attack variant
10 |   |   until Relieve_msg = TRUE or TokenReceived = FALSE;
11 |   until termination;
```

$Revert_{msg}$ in Algorithm 6.2. Similarly, the Algorithm 6.7 creates a condition for inputs to be manipulated instead of reverting the strategy as shown in Algorithm 6.3. Correctness proof is similar to the one given in 3.1.

6.4 Classifying Cheating Attacks

The *ACA* Framework described allows a variety of attacks which are beyond only cheating attacks to be designed. Overall, the set of attacks that can be designed at a time is governed by the intent of an attacker. For instance, cheating intent can be considered as a function that is applied by an attacker to the property dimension Pr Dimension of the Attack-Domain in order to influence profit distribution. The RCSMG platform warrants that such attacks be resource aware in order to be successful and to avoid easy detection. As such, a concise *Attack-Domain Model* will define the upper and lower bounds of the system resources in which plausible attacks can be executed.

Consider the decentralised CDA as the appropriate auction mechanism for power allocation in a smart grid. Such a decentralised CDA will determine the states, attack vectors and tentative procedures on which an attack can be build on. In turn, this implicitly narrows the set of plausible attacks that are likely to manifest such a CDA scheme. It is due to this reasoning that traditional cheating attacks can be eliminated as there are unlikely to manifest on the given CDA states and attack vectors. For example in a centralised auction the auctioneer is a vector for the attack. This is not necessarily the case in a decentralised scheme considered herein (as proposed in [5]). The attack model in turn specifies the attacker model with respect to the CDAs states and attack procedures, which are guided by the attacker's capability and rational. Cheating attacks can be classified with respect to the attacker's capabilities and attacker's rationality, in that order.

Attacks are categorised according to attacker capabilities in step 7 in Fig. 6.1: Limited, Advanced (A and B) and Expert (shown in Table 6.1). For the purpose of this chapter we will sideline Attacker B type and focus on Attacker A as the most likely capable attacker types to bring forth automated forms of cheating. At step 8 of the *ACA* Framework, the attackers can be further categorised using the Rational Model (Sect. 6.2.2.3) according to the number of adversaries and their interaction as follows: a single attacker; multiple non-cooperating attackers; and multiple cooperating attackers (Fig. 6.4). The single attacks can be carried out even by multiple adversaries, with an additional complexity on the impacts and the respective attack model that can be designed.

Table 6.1 Adversary types and capabilities

Adversary capabilities			
Attacker types	Knowledge	Disclosure	Disruption
1 Limited attacker	High	–	–
2 Advanced attacker A	High	Yes	–
3 Advanced attacker B	High	–	Yes
4 Expert attacker	High	Yes	Yes

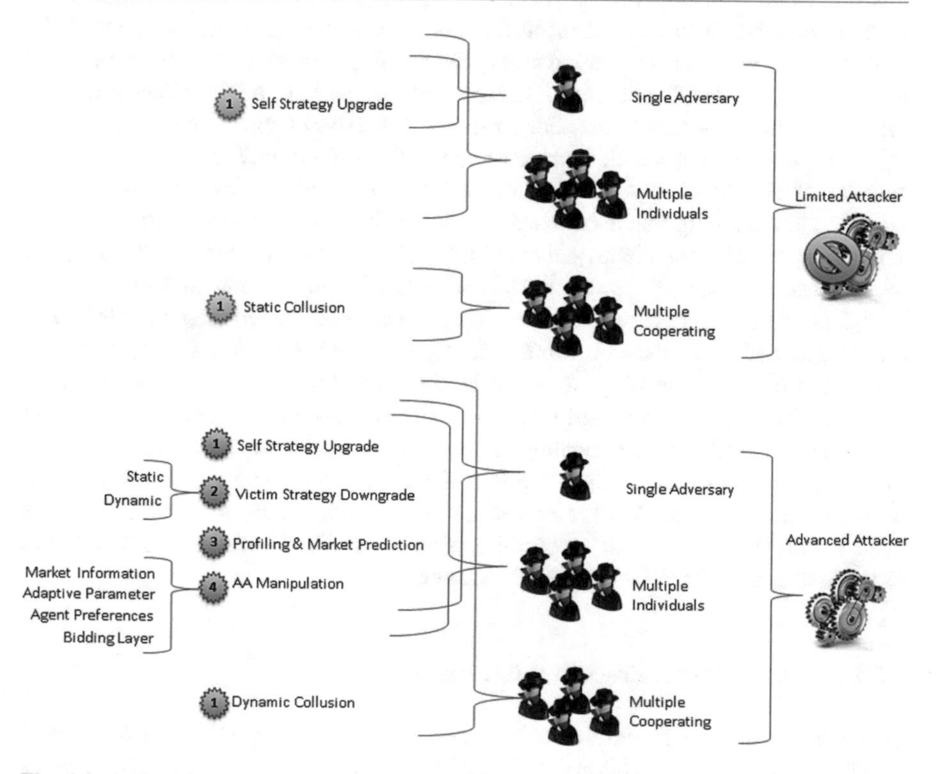

Fig. 6.4 A classification of cheating attacks on decentralised CDAs

6.5 Formulating Counter Measures

Attacks are designed to achieve a goal inherent to cheat. This implies some financial benefit is sort by the attacker, usually at the cost of system performance. Intuitively, one way in which a defender of a decentralised CDA system can address such forms of attacks is by detecting inconsistent behaviour such as profit distribution and system performance properties.

6.5.1 Exception Handling Approach

In [7, 8], cheating attacks are observed as giving rise to exceptions; situations which fall outside the normal operating conditions expected of the CDA and its components. As such, cheating attacks' exceptions are resolved through *exception handling* by distinct domain-independent agents- the citizen approach [62]. The authors proposed two properties of the CDA that can be use the exception handlers for positive detection and mitigation. The first property is allocative efficiency – a measure of how well the market runs. This measure is given as a ratio of the profit made during the auction to the profit that could be made if the agents traded in the most efficient way (if each offered at its private value, and the traders were matched to maximise the profits obtained). This provides a measure of the effectiveness of the market in economic terms. For instance, the experimental results motivating the collusion cheating attack (Sect. 6.3), show that colluding *TA*s may gain higher surplus while a decrease in the allocative efficiency will be observed. Intuitively, an exception handling mechanism should use such information to positively detect a cheating attack and identify the culprit. The second property is the number of messages exchanged by an individual *TA*. Any slight and sudden increase in the number of messages a *TA* exchanges will raise a red flag incident. The aforementioned two measures were recently used together to detect and resolve automated forms of cheating in the described work [7, 8]. The work employs an exception handling EH protocol that uses sentinels that monitor and collaborate in order to detect and mitigate cheating attacks. In this book chapter we consider how this approach can be used in solving the additionally proposed attacks. In order to integrate EH, the Local Market Procedure of the initial CDA Algorithm 6.1 have been modified and additional procedures that the sentinels will use to detect and mitigate cheating attacks have been developed.

6.5.1.1 Local Market Procedure Extension

Effectiveness of the decentralised CDA is based on its arguably efficient token-based MUTEX protocol. EH addition does not affect the MUTEX properties since it does not interfere with the token distribution process. The token-handling protocol is not altered except for additional time (*TA*s now require additional time) that should be included in the token handling timers. The initial constraints in our CDA algorithm are maintained.

Sketch Proof: The algorithm procedure gets inputs from the central sentinel in the variables: $TA_i Pen$ (a positively identified adversary TA to be penalised), $TA_i v$ (a positively identified victim TA). $TA_a[.]$ represents an array of all positively identified adversary TAs. When an TA_i receives a token (line 8), and a certain TA_is have been identified as adversaries (line 9), the *TA*s in question will be penalised (line 10) and the human participant alerted (line 11). Otherwise, the 'normal' procedure described in Algorithm 6.1 is executed.

Algorithm 6.8: $LocalMarketExecution$ Extension

```
   Input  : blacklist, TokenReceived, TA_a[.]
   Output : FlagM_mp, R, TokenCounter, TokenOB, LocalOB
1  Initialisation: FlagM_sm ← FALSE, TokenReceived ← FALSE, R ← 1000,
   TokenCounter ← 0, TA_i − Sentinel ← ∅ ;        // (where i = 0, 1, 2...N)
2  for each TA ∈ TA_i[.] do
3  |  if TokenRecieved = TRUE then
4  |  |  if TA_i ∈ blacklist and TA_i ∈ TA_a[] then
5  |  |  |  Penalise[TA_i] ;                       // penalised for cheating
6  |  |  |  Alert Participant! ;
7  |  |  else
8  |  |  |  TokenCounter ++ ; // the market execution continues as
   |  |  |  normal
9  |  |  |  Set FlagM_mp to TRUE ;
10 |  |  |  Ask TA_i-Sentinel for report ;
11 |  |  |  Ask TA_i to FormOffer() ;              // TAs submits its offer
12 |  |  |  Execute Trade() ;        // TAs offer is accepted or rejected
13 |  |  |  Update LocalOB from TokenOB ;
14 |  |  |  return TokenOB ;
15 |  else
16 |  |  wait for TOKEN ;
```

Sketch Proof: This sub-algorithm is similar to 6.1, with slight changes in lines 8–10 which capture the conditional statement allowing for the penalisation and notification of a cheating agent. If a *TA* is blacklisted and some *TA*s have been identified as victims, the blacklisted *TA* is penalised for cheating. Say, S **if c then** A_1 **else** A_2 **endif**, where c is well defined (meaning it can be evaluated), $P \wedge c \overset{A}{\to} Q$ and $P \wedge \bar{c} \overset{A}{\to} Q$, then $P \overset{A}{\to} Q$. We can verify the correctness of the branch (both when c is true and when c is false). Given the algorithm's precondition, P, is $TokenReceived = \{0, 1\}, blacklist = \{0, 1\}$ and the postcondition, Q, is a $Penalise[TA_i]$ or $TokenOB$ update, when the condition, c, that is $TA_i \in$

$blacklist$ and $TA_i \in TA_a[.]$ is satisfied, the $Penalise[TA_i]$ variable is set to TRUE and the participant is alerted. In the other case, the auctioning process goes as in Algorithm 6.1. This holds for any number of $TokenReceived$ or blacklisted *TA*s observed.

6.5.1.2 TA-Sentinel Execution

Consider sentinels (*TA*–Sentinels) to have no access to the internal state of the *TA*s their associated with as this could open up the system to a myriad array of adversaries through sentinel compromise. We assume *TA*–Sentinels and the associated *TA* reside on the same M_{mp} (similar to [7]). Each *TA*–Sentinel is capable of monitoring all messages exchanged by the associated *TA*s. Algorithm 6.9 presents a pseudocode of the *TA*–Sentinel procedure. Each *TA*–Sentinel knows the expected maximum number of messages, u, and some irregular messages, v, that can be exchanged by the associated *TA* (line 5). Thus, if a sentinel, $TA - Sentinel_1$, monitoring say TA_i, observes an anomaly of $(u + v)$ messages it will red-flag TA_i (line 7). Red-flagging is the first step in detecting an exception that allows the *TA*–Sentinel to keep records of this incident for future reference (Algorithm 6.9). On arrival of the token, when $FlagM_{mp}$ is set to TRUE, a *TA*–Sentinel will pass the report to the T–Sentinel.

Sketch Proof: The Algorithm 6.9 allows the TA–Sentinel to observe the messages exchanged by the *TA* it is monitoring. For each TA (line 2), the outer loop (line 3), guarantees the sub-algorithm terminates at the end of a trade day t_{round}. The inner loop (line 4) will always terminate when $FlagM_{mp}$ is TRUE (on the receipt of the TOKEN at the mobile phone M_{mp} hosting the TA). The loop ensures the TA–Sentinel listens to messages based on the condition that additional messages

Algorithm 6.9: $TA - SentinelExecution$ Procedure

```
   Input : FlagM_mp
   Output: RedFlag[TA_i]
1  Initialise: u = 0, RedFlag = FALSE
2  for each TA-Sentinel ∈ TA_i-Sentinel[.] do
3  |   repeat
4  |   |   repeat
5  |   |   |   Listen for messages ;
6  |   |   |   if messages > u then
7  |   |   |   |   set RedFlag = TRUE ;   // when additional messages are
   |   |   |   |   observed
8  |   |   |   else
9  |   |   |   |   set RedFlag = FALSE ;
10 |   |   until FlagM_mp = TRUE;
11 |   until Termination;
```

are observed. Say, S **if c then** A_1 **else** A_2 **endif**, where c is well defined (meaning it can be evaluated), $P \wedge c \stackrel{A}{\rightarrow} Q$ and $P \wedge \bar{c} \stackrel{A}{\rightarrow} Q$, then $P \stackrel{A}{\rightarrow} Q$. We can verify the correctness of the branch (both when c is true and when c is false). Given the algorithm's precondition, P, is $FlagMmp = \{0, 1\}$ and the postcondition, Q, is a $RedFlag[TA_i]$, when the condition, c, is $messages > u$ is satisfied, the $RedFlag$ variable is set to TRUE. In the other case, $Redflag$ is set to FALSE. This holds for any number of messages observed.

6.5.1.3 T–Sentinel Execution

We consider another special type of sentinel that is embedded into the mobile token (T–Sentinel). The token carries the auction order-book and respective records of the market, which include the current allocative efficiency and each TAs surplus. We assume the T–Sentinel has access to this information in order to analyse trade history to confirm cheating of *TA*s. Algorithm 6.10 presents the pseudocode that is executed by the T–Sentinel. The token is accessible to all *TA*s by virtue of the MUTEX protocol. On arrival at, say TA_i, the TA_i–Sentinel will be prompted to submit a report to the T–Sentinel on condition that a red-flag was recorded prior to token arrival. On reception of the report (line 6), the T–Sentinel will blacklist TA_i for further enquiry at the end of the trade day (line 7). The reason is cheating resolution can only be comprehensive if adequate market data has been collected and analysed. By incorporating the T–Sentinel into the order-book, our approach ensures: no additional messages are incurred. At the end of the trade day (line 10) the T–Sentinel will be tasked with ascertaining the cheating cases presented herein.

Sketch Proof: The Algorithm 6.10 executed by the T–Sentinel ensures receipt of *Redflags* (which leads to blacklisting of *TA*s) and the identification of the victims and adversary by considering the allocative efficiency, actual surplus and expected surplus. The first loop (line 2) ensures the red-flagged TAs that are identified are blacklisted until end of trading day t_{round}. Assume S, **if** c_1 **then** A_1 **elseif** c_2 **then** A_2 **else** A_3 **endif**, where c_1 and c_2 are well defined (meaning these can be evaluated), $P \wedge c_1 \stackrel{A}{\rightarrow} Q$ or $P \wedge c_2 \stackrel{A}{\rightarrow} Q$, then $P \stackrel{A}{\rightarrow} Q$. We can verify the correctness of the branches (when both c_1 and c_2 are true and when these are false). Given the algorithm's precondition, P, is the $RedFlag$ and the *AllocativeEfficency*, while the postcondition, Q, is a TA_iPen, TA_iVic confirmation or false flag indication. If the condition, c_1, is $TA_i \in blacklist$ **and** *ActualSurplus* > *ExpectedSurplus* while condition c_2 is $TA_i \in blacklisted$ **and** *ActualSurplus* < *ExpectedSurplus*

Algorithm 6.10: $T - SentinelExecution$ Procedure

Input : $RedFlag[TA_i]$, $ExpectedSurplus$, $ExpectedEfficiency$
Output: $TA_i Pen$, $TA_i Vic$

1 Initialise: $RedFlag = FALSE, blacklist = \emptyset, TA_i Pen = \emptyset, TA_i Vic = \emptyset$
2 **repeat**
3 | Receive of $RedFlag$;
4 | $blacklist \leftarrow RedFlag[TA_i]$;
5 **until** $termination$;
6 Calculate Allocative Efficiency ;
7 **if** $blacklist \neq \emptyset$ ***and*** *AllocativeEfficiency* < *ExpectedEfficiency* **then**
8 | Cheating Confirmed! ;
9 | **if** $TA_i \in blacklist$ ***and*** *ActualSurplus* > *ExpectedSurplus* **then**
10 | | TA_i is an adversary! ;
11 | | $TA_i Pen \leftarrow TA_i$; `// The TA is identified as an adversary`
12 | **else if** $TA_i \in blacklisted$ ***and*** *ActualSurplus* < *ExpectedSurplus* **then**
13 | | TA_i is a victim! ;
14 | | $TA_i Vic \leftarrow TA_i$; `// The TA is identified as a victim`
15 | **else**
16 | | False flag!;
17 **else**
18 | No Cheating! ;

are satisfied, the $TA_i Pen$, $TA_i Vic$ can always be confirmed and S verified. In the other case, A false flag is reported due to the condition in line 7.

6.5.2 Cheating Detection and Mitigation

Specific *TA* surplus margins form the core requirement in distinguishing the cheating nodes from well behaving *TA*s. Sudden increase in surplus is the end goal of adversarial nodes, thus making it the best parameter to use in detecting cheating.

Strategy-Downgrade Cheating
If a red-flagged incident, of say, TA_i happens to coincide with other *TA*s red-flag incidents, agent manipulation or collusion can be re-affirmed with greater probability. Intuitively, its implied that extra messages are being exchanged by *TA*s, indicating plausible *TA* manipulation. Thus, *Strategy-Downgrade Cheating* is confirmed by:

- a sudden decrease in the market allocative efficiency;
- evidence of more than 1 blacklisted agent in previous trade rounds;
- a sudden decrease in the number of wins by red-flagged *TA*s;
- identification of an individual *TA* with a constantly higher surplus, while the other *TA*s have distinctly low surplus.

Dynamic Collusion Attack
Similar to the Strategy-Downgrade Attack, adversary *TA*s can positively be identified by the extra surplus they gain. One evident problem in this notion is the scheme would have to store the surplus margins of individual *TA*s for further inquiries. A possible workaround is ensuring accurate surplus records are made and kept by the T–Sentinel as soon as a successful trade is made. *Collusion Attack* is confirmed by:

- a sudden decrease in market allocative efficiency;
- a number of blacklisted agents in previous trade rounds;
- sudden increase in the number of wins by red-flagged *TA*s;
- identification of a subset of η *TA*s constantly obtaining a higher surplus in as many rounds.

Evasive Attack
The Evasive adversary can positively be identified by a significant gain in additional surplus and an unusual rise in a number of messages passed by the adversary agent. *Evasive Attack* is confirmed by:

- blacklisting of 1 or more agents in previous trade rounds;
- sudden increase in the number of wins by red-flagged *TA*s;

AA Strategy Manipulation Attack
Detection can follow the same notion described for the strategy downgrade attack. Thus, *AA Strategy Manipulation Attack* is confirmed by:

- a sudden decrease in the market allocative efficiency;
- blacklisting of 1 or more *TA*s in previous trade rounds;
- a sudden decrease in the number of wins by red-flagged *TA*s;
- identification of an individual *TA* with a constantly higher surplus, while the other *TA*s have distinctly low surplus.

Resolution will follow positive identification/ confirmation of cheating. If a blacklisted TA_i is considered as an adversary TA_a, it will be penalised in the next trade day. If a sufficiently large amount of penalty is imposed on a discovered cheating *TA*, cheating will not be profitable for the adversary. There are several methods to impose such a penalty. For example, a form of a security deposit similar to one described in [37] can be utilised. If a *TA* does not cheat, the security deposit would be returned and when caught cheating it would be confiscated. Similarly, the cheating *TA* can be disallowed from trading in the current round or subsequent rounds. Further, if a *TA* is penalised the human participant is alerted of their penalisation. In cases where the human contests the penalty, further investigation can be done by the system administrators for instance.

6.5.2.1 Performance Analysis

We use message and time complexity analysis to provide a sketch evaluation of the EH scheme:

Message Complexity
Overall, in constructing the *EH* scheme the emphasis in on ensuring that no significant additional message overheads are incurred. Thus, a slight deviation in the message complexity from the initial CDA algorithm was expected. At the M_{sm} level the *EH* solution involves an exchange of $\mathcal{O}(logN)$ messages in order to pass the token (orderbook) per market auction execution under light demand, per critical section execution. At the M_{mp} level, apart from $4n$ messages passed by *TA*s in order to execute, the only additional messages are those between the T- and *TA*–Sentinels: T–Sentinel prompting *TA* to report = 1; *TA*–Sentinel reporting = 1; T–Sentinel issuing a penalty = 1; resulting in $3n$ additional messages where n is the number of *TA*s. Total messages exchanged at the M_{mp} level would be $7n$. Overall message complexity expected in light demand is therefore $\mathcal{O}(nlogN)$.

Time Complexity
We consider the input of the overall algorithm to be M, that is the number of *TA*s participating in the CDA. Since operation in *T–SentinelExecution Procedure* is dominant of the two detection and resolution procedures, it is executed M times by the T–Sentinel. We expect the detection and resolution algorithms to run in linear time complexity $\mathcal{O}(M)$.

6.6 Related Work

Studies on CDA algorithms within grid-like platforms focus on resource allocation (see e.g. [10, 12, 19, 43–46]) and bidding strategies (see e.g., [18, 33]). But, security and cheating in CDAs has not been studied deeply perhaps because of the complexity associated with analysis of such auction schemes. To the best of our knowledge, the works of Wang and Leung [40], Trevathan, Hossein and Read [41] and recently Marufu et al [7, 8] are the only closest works that have tackled security issues related to CDAs. Wang and Leung [40] and Trevathan, Hossein and Read [41], tackle the issues of cheating in a centralised CDA from a defender's viewpoint. Both works outline preferred security properties or security goals that a "secure CDA" should satisfy. To guarantee the security goals and to deter cheating attacks aimed at violating privacy and anonymity of traders, a digital signature is incorporated. The cheating attacks considered in these works are similar in nature to the cheating attacks in single-sided auction protocols. Such an approach may be flawed in guaranteeing CDA defenders with the necessary information to provide detection and resolution measures. As pointed out by Marufu et al. in [7, 8], cheating is auction mechanism specific, which implies security properties and effective defence strategy also follow the same notion. Marufu et al. approach the issue of security in a decentralised agent-based CDA by modelling random cheating attacks, which can be helpful in understanding potential vulnerabilities and plausible exploits. However, such an approach results in specific attacks that may be difficult to generalise (since there are specific to the auction algorithm they target) and does not provide enough coverage of cheating attacks that can help defenders to develop more elegant and effective solutions. In order to provide such a broader coverage of attacks, one envisages the use of security quantification approach.

According to [53], security quantification can take one of three forms: analysis of large amount of logged operational data; use of simulation techniques and tools; and construction of analytic models. Analysis of logged data is straightforward, but less desirable. Such an analysis is practical, only useful after an incident of a security breach occurs. Additionally, it can be an expensive approach, as it requires building a real system, taking measurements and analysing the data statistically. Thus, simulation techniques and tools can provide an alternative, but they also suffer from lack of appropriate techniques and tools specific to security quantification. Constructing analytical models provides a better option as it can be performed in an a priori manner and is less costly. In this book chapter, we carry out a model-based security evaluation approach to provide coverage of attacks in a decentralised CDA. Such an approach is novel within the CDA security research community.

The existing model-based security evaluation approaches can be categorised from the attacker behaviour viewpoint into behavioural [53, 64, 65] and non-behavioural/ technical approaches [42, 66–68]. As observed in [53, 65], attack modelling approaches proposed in the literature have focussed mostly on the technical aspects and finding possible attack vectors. Technical approaches are therefore based upon simplistic assumptions about the factors that may affect attacker's decisions without considering factors such as the costs of bribing people that may affect an attacker's decisions on whether to perform the attack, and how to perform it. In this book chapter we propose an *ACA* Framework (Sect. 6.2) that provides more coverage of attacks on a decentralised CDA by considering both technical and behavioural/non-technical aspects of attack modelling. The framework is inspired by the work of Adepu and Mathur [42]. Adepu and Mathur developed a framework that enables researchers to design a variety of cyber and physical attacks for the assessment of attack detection methods and tools. Their framework defines attackers intent but fails to further explicitly consider human behaviour aspects (e.g. attacker's rational and attacker's capability) which our framework considers.

6.7 Conclusions

We present the *ACA* Framework that can be used to design cheating attacks on a CDA. The framework consists of a domain model, an attacker model, and an attack model. The attack-domain model of a CDA captures the elements that could serve as the target of an attacker. An attacker model captures intents as functions on the domain model to specify the target and intent of an attacker. The attack model captures the essential elements of an attack including: attack procedure; attack-domain; points-of-attack; attackers capability; attackers rational; and the start and end states of the attack. Our models in the framework are able to consider the rationality behind launching an attack (not formally) as well as the attacker's skills and knowledge of the CDA; unlike a similar framework that inspired this work [42]. In this chapter the broad applicability and utility of the framework is demonstrated by designing cheating attacks on decentralised CDAs for constrained smart micro-grids. The examples considered are definitely not exhaustive. We do not formally analyse the impact of each attack. This is part of the current work the authors are looking into to improve on the results of this chapter. More attacks that can be constructed from a different intent, for example of an attacker learning and experimenting with disruptive resources in his arsenal can also be interesting path for future work. We classify the cheating attacks in a decentralised CDA according to attacker's capabilities (limited and advanced attackers) and the number of attackers a particular attack can have (single, multiple coordinating, and multiple non-coordinating). As part of the proposed mitigation measures, an exception handling based solution is proposed. The solution is a tool for detection and resolution of some forms of cheating through the identification and reprimand of the culprit trader. This solution does not offer intrusion tolerance, which is, the ability of the system and its components to perform their intended function in spite of partially successful attacks. This can be an interesting direction that can be pursed as future work. We envisage that continued work in designing different types of attacks can provide a vast array of resources for system defenders to ensure CDA realise their fullest potential within resource constrained smart micro-grids.

Acknowledgements This work is part of the joint SANCOOP programme of the Norwegian Research Council and the South African National Research Foundation under NRF grant 237817; funded by Hasso-Plattner-Institute at UCT; and UCT postgraduate funding.

References

1. A. Kayem, C. Meinel, and S. D. Wolthusen, "A smart micro-grid architecture for resource constrained environments," in *In proceedings of the 2017 IEEE 31st International Conference on Advanced Information Networking and Applications, AINA*, (Taipeh, Taiwan), pp. 857–864, IEEE, Mar. 2017.

2. G. K. Weldehawaryat, P. L. Ambassa, A. M. C. Marufu, S. D. Wolthusen, and A. Kayem, "Decentralised scheduling of power consumption in micro-grids: Optimisation and security," in *Security of Industrial Control Systems and Cyber-Physical Systems - Second International Workshop, CyberICPS 2016, Heraklion, Crete, Greece, September 26–30, 2016, Revised Selected Papers*, vol. 10166 of *Lecture Notes in Computer Science*, (Heraklion, Greece), pp. 69–86, Springer, Sept. 2016.
3. P. L. Ambassa, A. Kayem, S. Wolthusen, and C. Meinel, "Secure and reliable power consumption monitoring in untrustworthy micro-grids," in *Future Network Systems and Security* (R. Doss, S. Piramuthu, and W. ZHOU, eds.), vol. 523 of *Communications in Computer and Information Science*, pp. 166–180, Springer International Publishing, 2015.
4. P. L. Ambassa, A. Kayem, S. D. Wolthusen, and C. Meinel, "Secure and reliable power consumption monitoring in untrustworthy micro-grids," in *International Conference on Future Network Systems and Security*, pp. 166–180, Springer, 2015.
5. A. M. C. Marufu, A. Kayem, and S. D. Wolthusen, "A distributed continuous double auction framework for resource constrained microgrids," in *ritical Information Infrastructures Security - 10th International Conference, CRITIS 2015, Berlin, Germany, October 5–7, 2015, Revised Selected Papers*, vol. 9578 of *Lecture Notes in Computer Science*, pp. 183–198, Springer, 2015.
6. A. M. C. Marufu, A. Kayem, and S. D. Wolthusen, "Fault-tolerant distributed continuous double auctioning on computationally constrained microgrids," in *Proceedings of the 2nd International Conference on Information Systems Security and Privacy, ICISSP 2016, Rome, Italy, February 19–21, 2016.*, pp. 448–456, SCITEPRESS, 2016.
7. A. M. C. Marufu, A. Kayem, and S. D. Wolthusen, "Power auctioning in resource constrained micro-grids: Cases of cheating," in *11th International Conference on Critical Information Infrastructures Security*, SPRINGER, 2016.
8. A. M. C. Marufu, A. Kayem, and S. D. Wolthusen, "Circumventing cheating on power auctioning in resource constrained micro-grids," in *2016 IEEE 18th International Conference on High Performance Computing and Communications; IEEE 14th International Conference on Smart City; IEEE 2nd International Conference on Data Science and Systems (HPCC/SmartCity/DSS)*, pp. 1380–1387, IEEE, Dec. 2016.
9. P. Pałka, W. Radziszewska, and Z. Nahorski, "Balancing electric power in a microgrid via programmable agents auctions," *Control and Cybernetics*, vol. 41, 2012.
10. J. Stańczak, W. Radziszewska, and Z. Nahorski, "Dynamic pricing and balancing mechanism for a microgrid electricity market," in *Intelligent Systems'2014 - Proceedings of the 7th IEEE International Conference Intelligent Systems IS'2014, September 24–26, 2014, Warsaw, Poland, Volume 2: Tools, Architectures, Systems, Applications*, pp. 793–806, Springer, 2015.
11. Z. Tan, *Market-Based Grid Resource Allocation Using A Stable Continuous Double Auction*. PhD thesis, The University of Manchester, 2007.
12. H. Izakian, A. Abraham, and B. T. Ladani, "An auction method for resource allocation in computational grids," *Future Generation Computer Systems*, vol. 26, no. 2, pp. 228–235, 2010.
13. A. Sobe and W. Elmenreich, "Smart microgrids: Overview and outlook," *CoRR*, vol. abs/1304.3944, 2013.
14. V. L. Smith, "An experimental study of competitive market behavior," *The Journal of Political Economy*, pp. 111–137, 1962.
15. C. Preist and M. van Tol, "Adaptive agents in a persistent shout double auction," in *Proceedings of the first international conference on Information and computation economies*, pp. 11–18, ACM, 1998.
16. S. H. Clearwater, R. Costanza, M. Dixon, and B. Schroeder, "Saving energy using market-based control," *Market-Based Control: A Paradigm for Distributed Resource Allocation*, pp. 253–273, 1996.
17. R. H. Guttman, P. Maes, A. Chavez, and D. Dreilinger, "Results from a multi-agent electronic marketplace experiment," in *Poster Proceedings of the Eighth European Workshop on Modeling Autonomous Agents in a Multi-Agent World (MAAMAW'97)*, pp. 86–98, Citeseer, 1997.
18. H. Ma and H.-F. Leung, "An adaptive attitude bidding strategy for agents in continuous double auctions," *Electronic Commerce Research and Applications*, vol. 6, no. 4, pp. 383–398, 2008.

19. P. Vytelingum, *The structure and behaviour of the Continuous Double Auction*. PhD thesis, University of Southampton, UK, 2006.
20. H. K. Nunna and S. Doolla, "Multiagent-based distributed-energy-resource management for intelligent microgrids," *IEEE Transactions on Industrial Electronics*, vol. 60, no. 4, pp. 1678–1687, 2013.
21. H. K. Nunna and S. Doolla, "Energy management in microgrids using demand response and distributed storage- a multiagent approach," *IEEE Transactions on Power Delivery*, vol. 28, no. 2, pp. 939–947, 2013.
22. H. K. Nunna, A. M. Saklani, A. Sesetti, S. Battula, S. Doolla, and D. Srinivasan, "Multi-agent based demand response management system for combined operation of smart microgrids," *Sustainable Energy, Grids and Networks*, vol. 6, pp. 25–34, 2016.
23. J. Rust, J. Miller, and R. Palmer, "Behavior of trading automata in a computerized double auction market," *The double auction market: Institutions, theories, and evidence*, pp. 155–198, 1993.
24. D. K. Gode and S. Sunder, "Allocative efficiency of markets with zero-intelligence traders: Market as a partial substitute for individual rationality," *Journal of political economy*, vol. 101, no. 1, pp. 119–137, 1993.
25. D. Cliff, "Minimal-intelligence agents for bargaining behaviors in market-based environments," *Hewlett-Packard Labs Technical Reports*, 1997.
26. D. Cliff, "Evolutionary optimization of parameter sets for adaptive software-agent traders in continuous double auction markets," *HP LABORATORIES TECHNICAL REPORT HPL*, no. 99, 2001.
27. S. Gjerstad and J. Dickhaut, "Price formation in double auctions," in *E-Commerce Agents, Marketplace Solutions, Security Issues, and Supply and Demand*, vol. 22, pp. 106–134, Elsevier, 2001.
28. G. Tesauro and R. Das, "High-performance bidding agents for the continuous double auction," in *Proceedings 3rd ACM Conference on Electronic Commerce (EC-2001), Tampa, Florida, USA, October 14–17, 2001*, pp. 206–209, ACM, 2001.
29. G. Tesauro and J. L. Bredin, "Strategic sequential bidding in auctions using dynamic programming," in *The First International Joint Conference on Autonomous Agents & Multiagent Systems, AAMAS 2002, July 15–19, 2002, Bologna, Italy, Proceedings*, pp. 591–598, ACM, ACM, 2002.
30. M. He, H. F. Leung, and N. R. Jennings, "A fuzzy-logic based bidding strategy for autonomous agents in continuous double auctions," *IEEE Transactions on Knowledge and data Engineering*, vol. 15, no. 6, pp. 1345–1363, 2003.
31. P. Vytelingum, R. K. Dash, E. David, and N. R. Jennings, "A risk-based bidding strategy for continuous double auctions," in *Proceedings of the 16th Eureopean Conference on Artificial Intelligence, ECAI'2004, including Prestigious Applicants of Intelligent Systems, PAIS 2004, Valencia, Spain, August 22–27, 2004*, vol. 16, pp. 79–83, IOS Press, 2004.
32. D. Cliff, "Zip60: an enhanced variant of the zip trading algorithm," in *The 3rd IEEE International Conference on E-Commerce Technology. The 8th IEEE International Conference on and Enterprise Computing, E-Commerce, and E-Services*, pp. 15–15, IEEE, 2006.
33. P. Vytelingum, S. D. Ramchurn, T. D. Voice, A. Rogers, and N. R. Jennings, "Trading agents for the smart electricity grid," in *9th International Conference on Autonomous Agents and Multiagent Systems (AAMAS 2010), Toronto, Canada, May 10–14, 2010, Volume 1–3*, pp. 897–904, International Foundation for Autonomous Agents and Multiagent Systems, 2010.
34. J. Trevathan, "Security, anonymity and trust in electronic auctions," *Crossroads*, vol. 11, pp. 2–2, May 2005.
35. J. Trevathan, *Privacy and security in online auctions*. PhD thesis, James Cook University, 2007.
36. J. Trevathan and W. Read, "Undesirable and fraudulent behaviour in online auctions.," in *SECRYPT 2006, Proceedings of the International Conference on Security and Cryptography, Setúbal, Portugal, August 7–10, 2006, SECRYPT is part of ICETE - The International Joint Conference on e-Business and Telecommunications*, vol. 6, pp. 450–458, INSTICC Press, 2006.

37. M. Yokoo, Y. Sakurai, and S. Matsubara, "The effect of false-name bids in combinatorial auctions: New fraud in internet auctions," *Games and Economic Behavior*, vol. 46, no. 1, pp. 174–188, 2004.
38. I. Chakraborty and G. Kosmopoulou, "Auctions with shill bidding," *Economic Theory*, vol. 24, no. 2, pp. 271–287, 2004.
39. R. Porter and Y. Shoham, "On cheating in sealed-bid auctions," *Decision Support Systems*, vol. 39, no. 1, pp. 41–54, 2005.
40. C. Wang and H. F. Leung, "Anonymity and security in continuous double auctions for internet retails market," in *System Sciences, 2004. Proceedings of the 37th Annual Hawaii International Conference on*, p. 10, IEEE, Jan. 2004.
41. J. Trevathan, H. Ghodosi, and W. Read, "An anonymous and secure continuous double auction scheme," in *System Sciences, 2006. HICSS'06. Proceedings of the 39th Annual Hawaii International Conference on*, vol. 6, p. 125, IEEE, 2006.
42. S. Adepu and A. Mathur, "Generalized attacker and attack models for cyber physical systems," in *40th IEEE Annual Computer Software and Applications Conference, COMPSAC 2016, Atlanta, GA, USA, June 10–14, 2016*, vol. 1, pp. 283–292, IEEE, 2016.
43. Z. Tan and J. R. Gurd, "Market-based grid resource allocation using a stable continuous double auction," in *Proceedings of the 8th IEEE/ACM International Conference on Grid Computing*, GRID '07, (Washington, DC, USA), pp. 283–290, IEEE Computer Society, 2007.
44. A. Haque, S. M. Alhashmi, and R. Parthiban, "Continuous double auction in grid computing: An agent based approach to maximize profit for providers," in *Proceedings of the 2010 IEEE/WIC/ACM International Conference on Intelligent Agent Technology, IAT 2010, Toronto, Canada, August 31 - September 3, 2010*, vol. 2, pp. 347–351, IEEE, 2010.
45. I. Koutsopoulos and G. Iosifidis, "Auction mechanisms for network resource allocation," in *Modeling and Optimization in Mobile, Ad Hoc and Wireless Networks (WiOpt), 2010 Proceedings of the 8th International Symposium on*, pp. 554–563, IEEE, 2010.
46. U. Kant and D. Grosu, "Double auction protocols for resource allocation in grids," in *International Symposium on Information Technology: Coding and Computing (ITCC 2005), Volume 1, 4–6 April 2005, Las Vegas, Nevada, USA*, pp. 366–371, 2005.
47. K. Raymond, "A tree-based algorithm for distributed mutual exclusion," *ACM Transactions on Computer Systems (TOCS)*, vol. 7, pp. 61–77, 1989.
48. Z. Zhang, F. Nait-Abdesselam, and P. H. Ho, "Boosting markov reward models for probabilistic security evaluation by characterizing behaviors of attacker and defender," in *Availability, Reliability and Security, 2008. ARES 08. Third International Conference on*, pp. 352–359, IEEE, 2008.
49. D. Gollmann, P. Gurikov, A. Isakov, M. Krotofil, J. Larsen, and A. Winnicki, "Cyber-physical systems security: Experimental analysis of a vinyl acetate monomer plant," in *Proceedings of the 1st ACM Workshop on Cyber-Physical System Security,CPSS 2015, Singapore, Republic of Singapore, April 14 - March 14, 2015*, pp. 1–12, ACM, 2015.
50. P. Liu, W. Zang, and M. Yu, "Incentive-based modeling and inference of attacker intent, objectives, and strategies," *ACM Transactions on Information and System Security (TISSEC)*, vol. 8, no. 1, pp. 78–118, 2005.
51. A. Teixeira, D. Pérez, H. Sandberg, and K. H. Johansson, "Attack models and scenarios for networked control systems," in *Proceedings of the 1st International Conference on High Confidence Networked Systems*, HiCoNS '12, pp. 55–64, ACM, 2012.
52. A. Teixeira, I. Shames, H. Sandberg, and K. H. Johansson, "A secure control framework for resource-limited adversaries," *Automatica*, vol. 51, pp. 135–148, 2015.
53. A. Sedaghatbaf and M. Abdollahi Azgomi, "Attack modelling and security evaluation based on stochastic activity networks," *Security and Communication Networks*, vol. 7, no. 4, pp. 714–737, 2014.
54. T. Amenaza, "Fundamentals of capabilities-based attack tree analysis," *Calgary, Canada, November*, 2005.
55. A. Buldas, P. Laud, J. Priisalu, M. Saarepera, and J. Willemson, "Rational choice of security measures via multi-parameter attack trees," *Critical Information Infrastructures Security*, pp. 235–248, 2006.

56. T. M. Chen, J. C. Sanchez-Aarnoutse, and J. Buford, "Petri net modeling of cyber-physical attacks on smart grid," *IEEE Transactions on Smart Grid*, vol. 2, no. 4, pp. 741–749, 2011.
57. B. Chen, Z. Kalbarczyk, D. M. Nicol, W. H. Sanders, R. Tan, W. G. Temple, N. O. Tippenhauer, A. H. Vu, and D. K. Yau, "Go with the flow: Toward workflow-oriented security assessment," in *Proceedings of the 2013 workshop on New security paradigms workshop*, pp. 65–76, ACM, 2013.
58. S. Jajodia and S. Noel, "Advanced cyber attack modeling analysis and visualization," tech. rep., DTIC Document, 2010.
59. P. Vytelingum, D. Cliff, and N. R. Jennings, "Strategic bidding in continuous double auctions," *Artificial Intelligence*, vol. 172, no. 14, pp. 1700–1729, 2008.
60. M. De Luca and D. Cliff, "Human-agent auction interactions: Adaptive-aggressive agents dominate," in *IJCAI 2011, Proceedings of the 22nd International Joint Conference on Artificial Intelligence, Barcelona, Catalonia, Spain, July 16–22*, vol. 22, p. 178, Citeseer, 2011.
61. D. Vach and M. A. Marsnales, "Comparison of double auction bidding strategies for automated trading agents," Master's thesis, Charles University in Prague, 2015.
62. P. Chaudhuri and T. Edward, "An algorithm for k-mutual exclusion in decentralized systems," *Computer Communications*, vol. 31, no. 14, pp. 3223–3235, 2008.
63. A. S. Kyle, "Continuous auctions and insider trading," *Econometrica: Journal of the Econometric Society*, pp. 1315–1335, 1985.
64. K. Sallhammar, S. J. Knapskog, and B. E. Helvik, "Using stochastic game theory to compute the expected behavior of attackers," in *2005 IEEE/IPSJ International Symposium on Applications and the Internet Workshops (SAINT 2005 Workshops), 31 January - 4 February 2005, Trento, Italy*, pp. 102–105, IEEE, 2005.
65. M. Niitsoo, "Optimal adversary behavior for the serial model of financial attack trees.," in *Advances in Information and Computer Security - 5th International Workshop on Security, IWSEC 2010, Kobe, Japan, November 22–24, 2010. Proceedings*, pp. 354–370, Springer, 2010.
66. B. B. Madan and K. S. Trivedi, "Security modeling and quantification of intrusion tolerant systems using attack-response graph," *Journal of High Speed Networks*, vol. 13, no. 4, pp. 297–308, 2004.
67. O. M. Dahl and S. D. Wolthusen, "Modeling and execution of complex attack scenarios using interval timed colored petri nets," in *Proceedings of the 4th IEEE International Workshop on Information Assurance (IWIA 2006), 13–14 April 2006, Egham, Surrey, UK*, p. 12, IEEE, 2006.
68. M. Kiviharju, T. Venäläinen, and S. Kinnunen, "Towards modelling information security with key-challenge petri nets," in *Nordic Conference on Secure IT Systems*, pp. 190–206, Springer, 2009.

Chapter 7
Inferring Private User Behaviour Based on Information Leakage

Pacome L. Ambassa, Anne V. D. M. Kayem, Stephen D. Wolthusen, and Christoph Meinel

Abstract In rural/remote areas, resource constrained smart micro-grid (RCSMG) architectures can provide a cost-effective power supply alternative in cases when connectivity to the national power grid is impeded by factors such as load shedding. RCSMG architectures can be designed to handle communications over a distributed lossy network in order to minimise operation costs. However, due to the unreliable nature of lossy networks communication data can be distorted by noise additions that alter the veracity of the data. In this chapter, we consider cases in which an adversary who is internal to the RCSMG, deliberately distorts communicated data to gain an unfair advantage over the RCSMG's users. The adversary's goal is to mask malicious data manipulations as distortions due to additive noise due to communication channel unreliability. Distinguishing malicious data distortions from benign distortions is important in ensuring trustworthiness of the RCSMG. Perturbation data anonymisation algorithms can be used to alter transmitted data to ensure that adversarial manipulation of the data reveals no information that the adversary can take advantage of. However, because existing data perturbation anonymisation algorithms operate by using additive noise to anonymise data, using these algorithms in the RCSMG context is challenging. This is due to the fact that distinguishing benign noise additions from malicious noise additions is a difficult

P. L. Ambassa (✉)
Department of Computer Science, University of Cape Town, Rondebosch, Cape Town, South Africa
e-mail: pambassa@cs.uct.ac.za

A. V. D. M. Kayem · C. Meinel
Hasso-Plattner-Institute, Faculty of Digital Engineering, University of Potsdam, Potsdam, Germany
e-mail: christoph.meinel@hpi.uni-potsdam.de

S. D. Wolthusen
Department of Mathematics and Information Security, Royal Holloway, University of London, Egham, Surrey, UK

Norwegian Information Security Laboratory, Gjovik University College, Norwegian University of Science and Technology, Trondheim, Norway
e-mail: stephen.wolthusen@rhul.ac.uk

A. V. D. M. Kayem et al. (eds.), *Smart Micro-Grid Systems Security and Privacy*, Advances in Information Security 71, https://doi.org/10.1007/978-3-319-91427-5_7

problem. In this chapter, we present a brief survey of cases of privacy violations due to inferences drawn from observed power consumption patterns in RCSMGs centred on inference, and propose a method of mitigating these risks. The lesson here is that while RCSMGs give users more control over power management and distribution, good anonymisation is essential to protecting personal information on RCSMGs.

Keywords Approximation algorithms · Electrical products · Home appliances · Load modeling · Monitoring · Power demand · Wireless sensor networks · Distributed snapshot algorithm · Micro-grid networks · Monitoring · Power consumption characterization · Sensor networks

7.1 Introduction

Resource constrained smart micro-grids (RCSMGs) describe a power management concept in which rural/remote communities are powered with an amorphous distributed formation of electricity generation devices, coordinated via a lossy network [1]. Applications can be found in housing estates, suburban localities, and academic or public communities such as universities or schools. Typical implementation models operate either in connected or island mode. In the connected mode the RCSMG is connected to the power grid and only becomes operational if, and when the grid is unable to power the affected area. The island mode by contrast, operates as an autonomous powering system, that is completely disjoint from the grid. We focus on the latter in this chapter, as a means of provisioning power to rural/remote areas.

7.1.1 Context and Motivation

As presented in Chap. 5, RCSMG can be designed as a self-contained energy-based cyber-physical system (CPS), composed of a physical system (Power Network) that is coordinated by a communication network. The communication network serves as platform for algorithmic control of the micro-grid. Power is generated from renewable energy sources, usually contributed by grid users. Since the generation units are distributed, the control system relies on distributed algorithms for power sharing. As a cost effective measure, the control system uses an information and communications infrastructure (cyber-layer) that is based a combination of low-cost computing and ubiquitous devices such as sensors, mobile devices, and wireless communication protocols. The inter-dependency between the cyber-layer and the physical layer (power network) introduces privacy vulnerabilities that can be exploited to subvert the operation of the RCSMG. For example, energy theft attacks can be provoked by observing usage behaviours, and based on this information, substituting legitimate power consumption data transmissions with false measurements. Recent work [2–4] has also demonstrated that centralised Smart Grids are vulnerable to privacy violations that target state estimation, energy marketing, and topological modifications. On RCSMGs detecting risks of privacy violations due to inferences drawn from observed power consumption patterns is determined by how accurately one can distinguish benign from malicious data distortions.

7.1.2 Problem Statement

Privacy preserving algorithms are designed to operate on the principle of modifying or deliberately distorting data to conceal sensitive information while maintaining utility. The goal is have a cost-benefit trade-off between data transformation and data utility. Maintaining data utility is useful for grid operations such as power scheduling and billing, but at the same time privacy is important in protecting users from targeted advertising, for instance. We are faced with two problems in preventing privacy leaks on RCSMGs:

1. Differentiating data distortions due to the conditions of the underlying network (e.g. bursty transmissions, component failures, modifications in network topology,. . .), from legitimate data perturbations.
2. Detecting malicious data manipulations and circumvention strategies.

7.1.3 Contributions

In this Chapter, we present some cases of privacy violations due to inferences drawn from observed power consumption patterns that emerge in RCSMGs. Specifically, we present examples to show how user behaviours can be deduced by observation of power usage profiles. As a next step we present an overview of a method of anonymising the data before it is transmitted. Our solution proposes a model for differentiating benign data distortions due to the underlying network from malicious distortions.

Outline of the Chapter

The remainder of the chapter is organised as follows. In Sect. 7.2, we present a model for a RCSMG as a basis for the examples of privacy violation risks discussed in Sect. 7.3. Section 7.4, presents a method of mitigating the presented cases of privacy violations due to inferences drawn from observed power consumption patterns due to malicious data distortions. Related work is discussed in Sect. 7.5. We offer concluding remarks in Sect. 7.6.

7.2 A Smart Micro-Grid Model

As mentioned before, we consider an island smart micro-grid model, composed of households, generation units, and small factories/businesses, for instance. Each component is a producer-consumer (prosumer) capable of consuming and producing energy. A user usually controls a component, and can assume either one of three roles namely, consumers, producers, and trading operators. Each generating component consumes a certain portion of the energy it generates. The surplus is sold to the network, but when extra energy is required, the user either sends out a broadcast or bidding request on a per-need or per-importance basis. In the following, subsections we provide an overview of the operation of the RCSMG architecture [1].

7.2.1 Notation

Household appliances are represented as a set $\mathscr{A} = \left\{a_1, \ldots, a_i, \ldots, a_{|\mathscr{A}|}\right\}$ where $|\mathscr{A}| \geq 1$ and a_i represents the ith appliance in the household. $|\mathscr{A}|$ is the cardinality of the maximum number of household appliances that can be linked to a household data aggregation unit $\mathscr{M}$. In each household there is at least one data aggregation unit $\mathscr{M}$ and consumption/generation data is collected at periodic intervals $\Delta t \in \mathscr{T} = 1, 2, \ldots, T$ where T is finite. Further details on power data characterisations on RCSMGs are provided by Ambassa et al. [5]. A set of authorised users $\mathscr{U}$ control $\mathscr{M}$. $\mathscr{U}$ is composed of subsets u_g, u_s, and u_r where u_g denotes users reporting power generation data, u_s users with power storage units, and u_r users providing power consumption reports. A user can belong both to u_g and u_s, but users in u_r cannot be in either one of, or both u_g and u_s. If $u_i \in \left\{u_g \wedge u_s\right\} \vee \left\{u_g \vee u_s\right\}$ then $u_i \notin u_r$ and if $u_i \in u_r$ then $u_i \notin \left\{u_g \vee u_s\right\}$. Each RCSMG consists of a set of C clusters of residential households and small businesses denoted by $C = \{c_1, \ldots, c_j, \ldots, c_N\}$ where c_j represents the jth cluster and N the maximum number of household clusters sustainable on the RCSMG. A cluster c_j is associated to a smart meter $\mathscr{SM}(j)$. A household is associated to one and only one cluster, $c_j \in C$ and linked to a smart meter ($\mathscr{SM}$). Each $\mathscr{SM}$ has a maximum nodal degree of d where d indicates the maximum number of households that $\mathscr{SM}$ can handle efficiently. We denote a household $h_{j,k}$ to indicate the kth household, where $1 \leq k \leq |d|$, in $c_j = h_{j,1}, \ldots, h_{j,i}, \ldots, h_{j,|d|}$ connected to $\mathscr{SM}(j)$ where $j \geq 1$ and $1 \leq j \leq N$.

7.2.2 Power Network

The power supply network is based on a prosumer model. We define $P_{h_{i,i}}$ as the power generated at $h_{i,i}$, and expressed as:

$$P_{h_{i,i}} = Gen_{h_{i,i}} - Load_{h_{i,i}}$$

where $Gen_{h_{i,i}}$ and $Load_{h_{i,i}}$ are the generation amount and the load respectively at $h_{i,i}$. $P_{h_{i,i}}$ is used to evaluate when a household is either a power consumer or producer. We can distinguish two scenarios: in the first, if $P_{h_{i,i}} > 0$ then $h_{i,i}$ has more generated electrical energy than necessary to satisfy its local demand. Second in the case of $P_{h_{i,i}} < 0$, $h_{i,i}$ is in need of electrical power.

A consumption estimate is provided by the household, and based on a power data model [5], the mobile device $\mathscr{M}$ computes a consumption estimate for billing. But, for simplicity, we focus on sensor supported reporting. For security reasons, each household declares a single $\mathscr{M}$, and a set of users $\mathscr{U}$ who are authorised to control $\mathscr{M}$. $\mathscr{M}$ can only report consumption or generation data from the household to which it is associated. For simplicity, only one user can make reports and the report are make asynchronously meaning that they not happen at a fixed period Δt. All authorisation changes must be explicit.

Each household cluster c_j linked to a shared smart meter $\mathscr{SM}_j$. Sharing smart meters is a cost-effective solution for deploying smart micro-grids in economically challenged areas. Household power consumption and/or generation reports are transmitted from the mobile devices ($\mathscr{M}$) to the smart meter ($\mathscr{SM}$) where they are aggregated for billing purposes.

Distributed Energy Resources (DER) handle power generation, while storage units such as batteries hold excess unconsumed power, and the physical delivery of electrical power to consumers is conveyed over standard power lines. Generated power is made available, at a cost, to grid participants and the excess generated power is stored in anticipation of future demand. The power network operates on a prosumer model, in which a household can be both a power producer (when a generator is attached to the household) and/or a power consumer. There are two possible power network configurations that we envisage.

The $\mathscr{SM}$ associated with a cluster say, c_j, will periodically advertise, based on network observations power availability. This information is transmitted to a utility provider $\mathscr{SM}$ who distributes the power according to need. Excess power is stored at units that have been advertised on the resource constrained smart micro-grid.

The set of electrical appliances $\mathscr{A} \geq 1$, in each household $h_{j,k}$ are each bound to a sensor $S(a_i)$ where $S(a_i)$ is the sensor associated with appliance a_i. The consumption or generation report for $S(a_i)$ over a period $\mathscr{T}$ is transmitted to the aggregation (mobile) device $\mathscr{M}$ where it is stored using a vector representation. In the consumption report vector, $S(a_i) \in \mathscr{M}\left(h_{j,k}\right)$ represents the value read from a_i during Δt for $h_{j,k}$, and is expressed as follows:

$$\mathscr{M}\left(h_{j,k}\right) = \left[S\left(a_1\right), \ldots, S\left(a_i\right), \ldots, S\left(a_{|\mathscr{A}|}\right)\right]$$

In this data structure each slot contains the value transmitted from $S(a_i)$ where $1 \leq i \leq |\mathscr{A}|$. When it is necessary to bound the number of household electrical appliances, and we note that $|\mathscr{A}| \leq |\mathscr{MAX}|$ where $|\mathscr{MAX}|$ is the maximum number of appliances that households in a given cluster can have. Reports from all $\mathscr{M} \in c_j$ are transmitted to $\mathscr{SM}_j$ where they are represented in the form of a two dimensional matrix $|d| \times |\mathscr{A}|$ where each row in the matrix represents the power system state for all $h_{j,k}$ during $\mathscr{T}$.

We now describe the communication network to show how the data is transmitted from the power network to the utility provider or to the coordinating smart meter.

7.2.3 Communication Network

The communication network is built on low-cost, low-processing computational devices and is comprised of three inter-dependent core sub-networks namely, the *Home Area Network (HAN)*, *Neighbourhood Network (NEN)*, and *Micro-Grid Network (MGN)*, that together form a multi-layer radio network. The entire network operates as an asynchronous distributed system with unreliable communication channels and untrustworthy nodes. Therefore, nodes can report false values either due to inherent faults or due to deliberate malicious manipulations. We discuss attack possible attack scenarios in Section III.

The HAN is represented by a household $h_{j,k} \in c_j$. In $h_{j,k}$ the $S(a_i)$ supporting the household appliances are organised to form a wireless sensor network whose communications are coordinated via a wireless sensor network communication protocol such as Bluetooth, ZigBee, and WiFi. This is because communication range between the household devices is a distance of roughly 10–100 m (line-of-sight). The HAN communicates both generated and consumed power to the NEN via the household's $\mathscr{M}$ which is typically a mobile device such as a cell phone.

The NEN sits outside the home area network, and consists of the cluster of homes c_j that are linked to the metering unit (shared smart meter), $SM(j)$. The SM is connected via a mesh network to all the $\mathscr{M}$ in c_j and the mesh network is supported by wireless communication protocols such as ZigBee, and WiFi which offer longer communication ranges, and are not impeded by obstructions.

Finally the MGN is the point where all of the information from the HAN and NEN is aggregated. We refer to this point as the micro-grid control centre (e.g. coordinating smart meter or utility provider) and use communication protocols, such as WiMAX, Cognitive Radio (CR) or 4G together with a wireless mesh topology to support data transmission from the NEN. HAN to NEN, and NEN to MGN communications are bilateral so, the communication network controls the power network, conveying messages reporting power generation/consumption from users to the grid coordinator and state estimation information is conveyed from the grid coordinator to the users/households. This is important in providing feedback on all levels of the grid thereby contributing to ensure trust, and performance reliability of the micro-grid. We now discuss how the control network handles power consumption, generation, and state estimation information to ensure causality in operations between the distributed network components.

7.2.4 Control Network

The control network facilitates automation of the micro-grid and is modelled as an acyclic graph $G = (\mathcal{SM}, E)$, where $\mathcal{SM}$ is a finite set of smart meters and E is a finite set of edges or communication paths (supported by the communication network—Section II C) between the smart meters. Each $\mathcal{SM}$ roots a tree comprised of household grouping devices $\mathcal{M}$, and sensors $S(a_i)$. The maximum number of descendent nodes ($\mathcal{M}$) that a $\mathcal{SM}$ can have, is bounded by $|d|$ to account for the physical capacity limitations of the $\mathcal{SM}$, and to ensure optimal reporting performance. In addition, each $\mathcal{M}$ has a maximum nodal degree of $|\mathcal{A}|$, which is the maximum number of appliances (represented by $S(a_i)$) that a household can link to $\mathcal{M}$. This is aimed at modelling a realistic scenario where for efficiency and monitoring purposes, only a given number of sensor nodes can be connected to any given $\mathcal{M}$. Each household, is represented on the tree schema by $\mathcal{M}$.

In the initial phase, the coordinating smart meter ($\mathcal{SM}_{old}$) is randomly selected and the duration period T_{op} agreed on. On termination of the period T_{op}, a leader election algorithm [6] is initiated by $\mathcal{SM}_{old}$ to decide on the next coordinating smart meter ($\mathcal{SM}_{new}$). The discussion of how the leader election algorithm operates is outside the scope of this paper, but the reader may refer to standard texts on distributed leader election algorithms for more information [7]. The report time intervals Δt is such that $\Delta t \in T_{op}$. Once the $\mathcal{SM}_{new}$ is agreed on, $\mathcal{SM}_{old}$ transfers all existing state estimation data to $\mathcal{SM}_{new}$. Using the scheme described by Ambassa et al. [8], the $\mathcal{SM}_{new}$ initiates power consumption/generation data collection, from the $N - 1$ other $\mathcal{SM}$s. In turn, each $\mathcal{SM}$ communicates to all its associated $\mathcal{M}$ to request the data. On reception of this message, the $S(a_i)$ linked to $\mathcal{M}$ collect the required data which $\mathcal{M}$ aggregates and transmits to $\mathcal{SM}$ from where the data is subsequently transmitted to $\mathcal{SM}_{new}$. The power consumption data is represented at $\mathcal{SM}_{new}$ using a three dimensional matrix data structure $N \times |d| \times |\mathcal{A}|$ as shown in Eq. (7.1).

$$\mathcal{SM}_{new} = \left(\mathcal{SM}_j\right) \begin{pmatrix} h_{j,1} \\ \dots \\ h_{j,k} \\ \dots \\ h_{j,|d|} \end{pmatrix} \begin{pmatrix} S(a_1) \\ \dots \\ S(a_i) \\ \dots \\ S(a_{|\mathcal{A}|}) \end{pmatrix} \tag{7.1}$$

where $j \geq 1$ and $j \leq N$.

7.3 Inferring Private User Behaviour Based on Consumption Data

The typical adversarial model for privacy in smart-grids follows a semi-honest model [9–12]. In this section, we extend this semi-honest privacy adversarial model to consider user behaviour. As a running example, we consider a scenario in which the adversary is internal to the RCSMG. The adversary can be either a user, smart meter or coordinator; and could be either *semi-honest* or *malicious*. In the *semi-honest* or *honest-but-curious* model, the adversary follows the execution protocol but, is curious to analyse data to deduce sensitive information about the identities of the owners of active electrical appliances as well as the type of appliances being used. On the other hand, In the *dishonest* or *malicious* model, the adversaries play an active role in interfering with the execution protocol. The malicious adversaries can also collude with each other to manipulate the data transmitted by compromising target components, and then use the compromised components to provoke privacy violations in the RCSMG. In the following we discuss cases of privacy violations due to inferences centred on observed power consumption patterns.

7.3.1 Inferential Attack Models

An inferential attack occurs in the context of RCSMG when a user gains information about the system either legitimately or maliciously. The user then proceeds to analyse the information to deduce sensitive information about electrical appliance usage, the presence, and behaviour of inhabitants in the target users' household. We consider two cases of inferential attacks namely, passive and active inferential attacks.

7.3.1.1 Passive Inference Attacks

Passive attacks can be provoked by an authorised adversary (Honest-but-curious) or a curious eavesdropper receiving or intercepting the load data transmitted by the targeted consumer $h_{j,k}$. We consider two categories of passive inference attacks namely, direct and indirect inferential attacks.

- **Direct inferential attack (DIA)** is the most simple form of attack. DIA uses energy data directly accessed, together with basic observations, to deduce some sensitive information. The DIA attacker seeks to gain information from the network traffic in the overlapping network coverage area. This information is analysed to infer activities in the target household. For instance, the sensors $S(a_i)$, report their measurements over a period t to a data aggregation unit, $\mathscr{M}$, in the neighbouring household. In this case, we assume that the $\mathscr{M}$ for household $h_{i,k}$, is not operational and so, $h_{i,k}$'s appliances must report consumption/generation to $\mathscr{M}\left(h_{j,k}\right)$. The adversary $h_{j,k}^{adv}$, intercepts these values due to the lossy communication medium and attempts to identify which appliances are in use in $h_{i,k}$. We assume that the adversary is aware of appliances' power consumption signature, as described in [8]. The intercepted and signature values are compared using a metric such as the Euclidean distance. The distance between the intercepted sensor values $S(a_i)$, and signature generated sensor values $\hat{S}(a_i)$ is computed as follows:

$$Dist(S(a_i), \hat{S}(a_i)) = \sqrt{\sum_{p=1}^{n}(S(a_i)_p - \hat{S}(a_i)_p)^2}.$$

Eliminating noise, when $d(S(a_i), \hat{S}(a_i)) = 0$), the corresponding appliance class is correctly identified. Based on correlations between observed appliance usage consumption reports and the household's activities, the adversary can further infer activities in the targeted household.

- **Indirect inferential attack (IIA)** aims to analyse energy time series data to infer private information such as household consumption, user behaviours and lifestyle [2]. We consider a dynamic system where at every reporting cycle, the delayed snapshot algorithm [8] collects, and computes the energy consumption $\mathscr{E}\left(h_{j,k}\right)$ of the household $h_{j,k} \in c_j$. In this case, $\mathscr{M}\left(h_{j,k}\right)$ reports the aggregated power consumption data for the period t, to $\mathscr{S}\mathscr{M}_j$. Here $\mathscr{S}\mathscr{M}_j$ is *honest-but-curious* in that $\mathscr{S}\mathscr{M}_j$ follows the execution protocol but uses the information collected to deduce sensitive information, in either one of the following ways:

  1. **Leakage from the power consumption data stream**. The adversary ($\mathscr{S}\mathscr{M}_j$) legitimately receives the household power consumption but would like to determine the appliances' usage patterns in the targeted household, i.e. when an appliance is ON or OFF [3]. To do this, the adversary $\mathscr{S}\mathscr{M}_j$ considers two consecutive power consumption reports. Let $\mathscr{E}\left(h_{j,k}\right)(t)$ be the power consumption reported from $h_{j,k}$ to $\mathscr{S}\mathscr{M}_j$ at time t and $\mathscr{E}\left(h_{j,k}\right)(t+1)$ be the one reported at time $t+1$. The difference between two consecutive reports can be used to identify step changes in appliance power consumption signatures. A step change occurs when $\left|\mathscr{E}\left(h_{j,k}\right)(t+1) - \mathscr{E}\left(h_{j,k}\right)(t)\right| > \alpha$ where α is determined by measurement noise. Based on the appliance's signature model [8], the adversary can distinguish load classifications by matching step changes with the appliance load signatures.
  2. **Leakage from a combination of power consumption and demand data streams**. Here, the model is similar to the previous case with the main difference being that household demand for future window cycles are considered. Here $h_{j,k}$ discloses his/her power requirements $\mathscr{D}\left(h_{j,k}\right)$ for the next scheduling cycle to $\mathscr{S}\mathscr{M}_j$ where $\mathscr{D}\left(h_{j,k}\right)^{\min}$ and $\mathscr{D}\left(h_{j,k}\right)_j^{\max}$ are the lower and upper bounds, respectively, of what the consumer is willing to accept in case of limited power availability. Based on the observed $\mathscr{S}\mathscr{M}_j$ values

reported, and aggregated at $\mathscr{SM}_{new}$ for power consumption scheduling, the adversary aims to infer the specific household appliances associated with a particular demand. Knowing the appliances' demand requirements can be used to define demand-inspired pricing mechanisms to raise profits and provision power unfairly. The sum of usage patterns is used to correlate multiple readings to improve the accuracy of the inference algorithm.

7.3.1.2 Active Inference Attacks

Passive inference attacks do not consider malicious interference with data transmissions. In active attacks by contrast, the attacker (malicious neighbours consumer or producer) actively manipulates data exchanges. Active attacks extend passive inferential attacks by injecting packets in the network. One example of an active attack is a data collusion attack, which we explain below.

In practice, before transmission of household power consumption, an anonymization method is applied to protect the user ID from directly linkability to the data. Therefore, we assume that explicit household identifiers are removed and replaced by a pseudonym before data transmission; and a cryptographic scheme is used to ensure that only authorised users received the data. We also assume that the adversaries have the capability of falsifying data and can collude with $\mathscr{SM}_j$.

To conduct an active attack, a dishonest neighbour say, $h_{j,k}^{adv}$, decides to distort power consumption data communicated over a given reporting period. Re-identification attacks illustrate this sort of attack by enabling the $\mathscr{SM}_j$ to re-identify the data stream (power consumption and demand) originating from the target household $h_{j,k}$. Data manipulations allow $\mathscr{SM}_j$ to easily link data originating from the same household $h_{j,k}$, making this attack much easier to conduct than presented in [13, 14].

7.4 Mitigating Privacy Inference

This section proposes a mitigation approach for privacy violations presented in Sect. 7.3. Our approach builds on the concept of differential privacy (DP) [15] which has been shown to be appropriate for SMG applications [10, 11, 16, 17]. In this case, the *honest-but-curious* $\mathscr{SM}_j$ can not infer individuals' data since sufficient noise is injected by DP to hide the individuals' data. DP is a state of the art privacy preserving approach based on a rigorous mathematical model. The DP approach consists of adding random noise to the power consumption/demand data reported by the consumer to conceal private consumer information [15]. (ϵ, δ)-differential privacy is formally expressed as follows:

Definition 7.1 (Differential Privacy [15]) A randomized function $\mathscr{S}: \mathscr{D} \mapsto \mathbb{R}^n$ satisfies (ϵ, δ)-differential privacy if for every pair of neighbouring data sets D_1 and

D_2, where D_1 and D_2 differ on at most one entry and for all ($M \subseteq Range(S)$),it holds that

$$P\left[S(D_1) \in M\right] \leq \exp(\epsilon) P\left[S(D_2) \in M\right] + \delta \tag{7.2}$$

Where M is a subset of the possible outputs of S and ϵ is a public privacy parameter that characterizes the level of privacy . A small value ϵ implies better privacy

In practice, DP can be achieved by simply adding a stochastic noise from a Laplace, Gaussian and exponential distribution. The Laplace mechanism works by introducing additive noise drawn from the Gaussian distribution, which is considered to be suitable for numeric data [18].

Laplace distribution $Lap(0, \lambda)$ with probability density function $f(x; \lambda)$), where the diversity parameter $\lambda = M(S)/\epsilon$, where $M(S) =^{max}_{D_1, D_2} \|S(D_1) - S(D_2)\|$ is the global sensitivity of the function S carefully calibrate to the global sensitivity of $\mathcal{S}$ [15].

While DP has been shown to work well in standard smart grid architectures [10, 11, 16, 17, 19], DP is not optimal for RCSMGs where the metering devices are low cost and the accuracy of the measurement limited [5, 20]. Adding noise to the data from the household affects the accuracy of the data which in turn causes a mismatch between reported and real energy consumption values. These issues can be avoided by defining a mechanism to add minimum noise while ensuring a certain level of privacy. Giraldo et al. [12] handle this problem in the control system by defining an *inherent differential privacy (IDP)* level that uses the inherent noise associated with the measurements to reduce the additional noise generated by the DP algorithm. In the following, we show how the inherent noise on lossy networks can be used to reduce the required additional noise for distortion by the DP algorithm, thereby minimising the negative impact on system utility and safety.

7.4.1 *Inherent Distributed Differential Privacy (IDDP)*

Our proposed scheme works as follows, the coordinator node $\mathcal{SM}_{new}$ begins by sending a query to the data aggregators ($\mathcal{M}$) requesting a power consumption or production or request report. The $\mathcal{M}$ in turn each query the set of associated sensors $S(a_i)$ linked to the household appliances a_i, and each reports its consumption/generation data back to $\mathcal{M}$ where the household power consumption/production/demand is computed. Each $\mathcal{M}$ adds noise to the data collected, and then encrypts the generated noisy data with a set of keys agreed on with the data collection node ($\mathcal{SM}_{new}$).Let $\bar{\mathcal{D}}(h_{j,k})$ be the noisy version of consumption/generation data at $\mathcal{M}\left(h_{j,k}\right)$ where $\bar{\mathcal{D}}(h_{j,k})(t) = \mathcal{D}(h_{j,k})(t) + \beta_k(t)$

and $\beta_k(t)$ is additive noise from the Gaussian distribution.[1] $\mathscr{M}\left(h_{j,k}\right)$ proceeds to send $\mathscr{D}(h_{j,k})(t)$ to the $\mathscr{S}\mathscr{M}_j$ from where $\mathscr{D}(h_{j,k})(t)$ is then transmitted to the coordinator node $\mathscr{S}\mathscr{M}_{new}$ for RCSMG operations such as power consumption scheduling. This process is repeated for the aggregated data from the set of households belonging to c_j and coordinated by $\mathscr{S}\mathscr{M}_j$. An aggregation protocol transmits the aggregated data to the coordinator node $\mathscr{S}\mathscr{M}_{new}$. The procedure is applied in a distributed manner so that the noisy household demand data $\bar{\mathscr{D}}(h_{j,k})$ is bounded by $\mathscr{D}(h_{j,k})(t)$. In this way, by using the property of Sequential composition of DP [21] we are able to enable distributed group privacy based on IDP from the household level instead of relying solely on $\mathscr{S}\mathscr{M}_j$, the intermediary metering device, to handle the IDP mechanism.

7.5 Related Work

Reliable access to energy is a key enabler for any modern society as electric power networks underpin most critical infrastructures and services. Rural/Remote areas that are not connected to national power networks, or that are subjected to load-shedding as grid operators prioritise urban centres in case of generator capacity shortages, are the most critically affected [22, 23, 23–27]. A key concern when deploying resource constrained smart micro-grids is power theft which often leads to overloading and grid destabilisation, but also discourages maintenance and user participation [28–30].

Privacy inference must also be addressed by controlling which portions of generation and load data are shared and with whom [19, 31]. Existing approaches consider the conventional centralised approach but this needs to be expanded to enable distributed and hierarchical privacy-preserving state estimation [32]. This is important when micro-grids are designed to rely on distributed energy generation source and where distributed control of the smart grid environments is necessary [23–25, 28, 33–37].

Existing work on privacy preserving techniques for SMGs includes data anonymization, data perturbation as well as cryptographic techniques. Cryptographic approaches such as homomorphic encryption protect consumer privacy by aggregating consumption data from several consumer without leaking private information [38, 39]. However, existing homomorphic encryption approaches are computationally expensive and incur a high processing and communication overhead [40], making them impractical for RCSMGs. Data perturbation approaches, like DP, are effective privacy preserving mechanisms in the area of smart power networks [10, 11]. On the one hand this approach theoretically preserves the privacy against arbitrary adversaries, but is vulnerable to power data correlation attacks [41]. In cyber-physical system Giraldo et al. [12]

[1] $\beta_k(t)$ is computed as per [12].

exploits a unique property of DP algorithms to design a tailored made DP with a minimum noise addition. This solution however, needs to be adapted to take into account the unreliable nature of RCSMGs.

7.6 Conclusion

We presented cases of privacy violations due to inferences drawn from observed power consumption patterns in RCSMGs. We considered abstractions of how the violations could potentially be provoked successfully based on a proposed RCSMG model. As a follow up, we presented a method of possibly circumventing these violations based on applying differential privacy to the power data in the case of indirect inference attacks.

References

1. A. Kayem, C. Meinel, and S. D.Wolthusen, "A smart micro-grid architecture for resource constrained environments," in *2017 IEEE 31st International Conference on Advanced Information Networking and Applications (AINA)*, March 2017, pp. 857–864.
2. M. Lisovich, D. Mulligan, and S. Wicker, "Inferring personal information from demand-response systems," *Security Privacy, IEEE*, vol. 8, no. 1, pp. 11–20, Jan 2010.
3. I. Rouf, H. Mustafa, M. Xu, W. Xu, R. Miller, and M. Gruteser, "Neighborhood watch: Security and privacy analysis of automatic meter reading systems," in *Proc. of the 2012 ACM Conf. on Comput. and Commun. Security*, ser. CCS '12. New York, NY, USA: ACM, 2012, pp. 462–473.
4. A. Molina-Markham, P. Shenoy, K. Fu, E. Cecchet, and D. Irwin, "Private memoirs of a smart meter," in *Proc. of the 2nd ACM Workshop on Embedded Sensing Systems for Energy-Efficiency in Building*, ser. BuildSys '10. New York, NY, USA: ACM, 2010, pp. 61–66.
5. P. L. Ambassa, A. Kayem, S. D. Wolthusen, and C. Meinel, "Secure and reliable power consumption monitoring in untrustworthy micro-grids," in *Future Network Systems and Security*, ser. Communications in Computer and Information Science, R. Doss, S. Piramuthu, and W. ZHOU, Eds. Springer International Publishing Switzerland: Springer International Publishing, 2015, vol. 523, pp. 166–180. [Online]. Available: http://dx.doi.org/10.1007/978-3-319-19210-9_12
6. M. Klonowski and D. Pajak, "Electing a leader in wireless networks quickly despite jamming," in *Proceedings of the 27th ACM Symposium on Parallelism in Algorithms and Architectures*, ser. SPAA '15. New York, NY, USA: ACM, 2015, pp. 304–312.
7. N. A. Lynch, *Distributed Algorithms*. San Francisco, CA, USA: Morgan Kaufmann Publishers Inc., 1996.
8. P. Ambassa, S. Wolthusen, A. Kayem, and C. Meinel, "Robust snapshot algorithm for power consumption monitoring in computationally constrained micro-grids," in *IEEE Innovative Smart Grid Technologies - Asia (ISGT ASIA)*, Nov 2015, pp. 1–6. [Online]. http://dx.doi.org/10.1109/ISGT-Asia.2015.7387160
9. S. McLaughlin, P. McDaniel, and W. Aiello, "Protecting consumer privacy from electric load monitoring," in *Proceedings of the 18th ACM Conference on Computer and Communications Security*, ser. CCS '11. New York, NY, USA: ACM, 2011, pp. 87–98. [Online]. Available: http://doi.acm.org/10.1145/2046707.2046720

10. E. Shi, T. H. H. Chan, E. G. Rieffel, R. Chow, and D. Song, "Privacy-preserving aggregation of time-series data," in *NDSS*, 2011.
11. G. Ács and C. Castelluccia, "I have a dream! (differentially private smart metering)," in *Inform. Hiding*, ser. Lecture Notes in Comput. Sci. Springer Berlin Heidelberg, 2011, vol. 6958, pp. 118–132.
12. J. Giraldo, A. Cardenas, and M. Kantarcioglu, "Leveraging unique cps properties to design better privacy-enhancing algorithms," in *Proceedings of the Hot Topics in Science of Security: Symposium and Bootcamp*, ser. HoTSoS. New York, NY, USA: ACM, 2017, pp. 1–12. [Online]. Available: http://doi.acm.org/10.1145/3055305.3055313
13. E. Buchmann, K. Böhm, T. Burghardt, and S. Kessler, "Re-identification of smart meter data," *Personal Ubiquitous Comput.*, vol. 17, no. 4, pp. 653–662, Apr. 2013.
14. V. Tudor, M. Almgren, and M. Papatriantafilou, "A study on data de-pseudonymization in the smart grid," in *Proceedings of the Eighth European Workshop on System Security*, ser. EuroSec '15. New York, NY, USA: ACM, 2015, pp. 2:1–2:6. [Online]. Available: http://doi.acm.org/10.1145/2751323.2751325
15. C. Dwork, "Differential privacy," in *33rd Int. Colloq. on Automata, Languages and Programming, part II (ICALP 2006)*, ser. Lecture Notes in Comput. Sci., vol. 4052. Venice, Italy: Springer Verlag, July 2006, pp. 1–12.
16. P. Barbosa, A. Brito, and H. Almeida, "A technique to provide differential privacy for appliance usage in smart metering," *Information Sciences*, vol. 370–371, pp. 355–367, 2016. [Online]. Available: http://www.sciencedirect.com/science/article/pii/S0020025516305862
17. V. Gulisano, V. Tudor, M. Almgren, and M. Papatriantafilou, "Bes: Differentially private and distributed event aggregation in advanced metering infrastructures," in *Proceedings of the 2Nd ACM International Workshop on Cyber-Physical System Security*, ser. CPSS '16. New York, NY, USA: ACM, 2016, pp. 59–69. [Online]. Available: http://doi.acm.org/10.1145/2899015.2899021
18. J. Cortés, G. E. Dullerud, S. Han, J. Le Ny, S. Mitra, and G. J. Pappas, "Differential privacy in control and network systems," in *2016 IEEE 55th Conference on Decision and Control (CDC)*, Dec 2016, pp. 4252–4272.
19. J. Giraldo, A. A. Cardenas, and M. Kantarcioglu, "Security vs. privacy: How integrity attacks can be masked by the noise of differential privacy," in *2017 American Control Conference (ACC)*, May 2017, pp. 1679–1684.
20. P. L. Ambassa, A. Kayem, S. D.Wolthusen, and C. Meinel, "Secure and reliable power consumption monitoring in untrustworthy micro-grids," in *Future Network Systems and Security*, ser. Commun. in Comput. and Inform. Sci., R. Doss, S. Piramuthu, and W. ZHOU, Eds. Springer, 2015, vol. 523, pp. 166–180.
21. C. Dwork and A. Roth, "The algorithmic foundations of differential privacy," *Found. Trends Theor. Comput. Sci.*, vol. 9, no. 3–4, pp. 211–407, Aug. 2014. [Online]. Available: http://dx.doi.org/10.1561/0400000042
22. Z. Wang and M. Lemmon, "Stability analysis of weak rural electrification microgrids with droop-controlled rotational and electronic distributed generators," in *2015 IEEE Power Energy Society General Meeting*. Piscataway, NJ, USA: IEEE Press, July 2015, pp. 1–5.
23. K. Iniewski, *Smart Grid: Infrastructure and Networking*. New York, NY, USA: McGraw Hill, 2013.
24. A. Gómez Expósito, A. Abur, A. de la Villa Jaén, and C. Gómez-Quiles, "A multilevel state estimation paradigm for smart grids," *Proceedings of the IEEE*, vol. 99, no. 6, pp. 952–976, jun 2011.
25. G. N. Korres, "A distributed multiarea state estimation," *IEEE Transactions on Power Systems*, vol. 26, no. 1, pp. 73–84, feb 2011.
26. R. Nagaraj, "Renewable energy based small hybrid power system for desalination applications in remote locations," in *2012 IEEE 5th India International Conference on Power Electronics (IICPE)*. Piscataway, NJ, USA: IEEE Press, Dec 2012, pp. 1–5.
27. E. D. Moe and A. P. Moe, "Off-grid power for small communities with renewable energy sources in rural guatemalan villages," in *Global Humanitarian Technology Conference (GHTC), 2011 IEEE*. Piscataway, NJ, USA: IEEE Press, Oct 2011, pp. 11–16.

28. Y. Feng, C. Foglietta, A. Baiocco, S. Panzieri, and S. D. Wolthusen, "Malicious false data injection in hierarchical electric power grid state estimation systems," in *Proceedings of the Fourth International Conference on Future Energy Systems*, ser. e-Energy '13. New York, NY, USA: ACM, 2013, pp. 183–192.
29. M. Perdue and R. Gottschalg, "Energy yields of small grid connected photovoltaic system: effects of component reliability and maintenance," *IET Renewable Power Generation*, vol. 9, no. 5, pp. 432–437, 2015.
30. P. Buchana and T. S. Ustun, "The role of microgrids amp; renewable energy in addressing sub-saharan africa's current and future energy needs," in *Renewable Energy Congress (IREC), 2015 6th International*. Sousse, Tunisia: IEEE Press, 24–26 March 2015, pp. 1–6.
31. A. Abur and A. Gómez Expósito, *Power System State Estimation: Theory and Implementation*. Boca Raton, FL, USA: CRC Press, 2004.
32. T. van Cutsem and M. Ribbens-Pavella, "Critical survey of hierarchical methods for state estimation of electric power systems," *IEEE Transactions on Power Apparatus and Systems*, vol. PAS-102, no. 10, pp. 3415–3424, oct 1983.
33. C. Gómez-Quiles, A. de la Villa Jaén, and A. Gómez Expósito, "A factorized approach to wls state estimation," *IEEE Transactions on Power Systems*, vol. 26, no. 3, pp. 1724–1732, aug 2011.
34. A. Gómez Expósito, A. Abur, A. de la Villa Jaén, C. Gómez-Quiles, P. Rousseaux, and T. van Cutsem, "A taxonomy of multilevel state estimation methods," *Electric Power Systems Research*, vol. 81, no. 4, pp. 1060–1069, apr 2011.
35. G. M. Mathews, "An optimal hierarchical algorithm for factored nonlinear weighted least squares state estimation," in *Proceedings of the 2012, 3rd IEEE PES International Conference on Innovative Smart Grid Technologies (ISGT Europe 2012)*. Berlin, Germany: IEEE Press, oct 2012, pp. 1–6.
36. A. Baiocco and S. D. Wolthusen, "Dynamic forced partitioning of robust hierarchical state estimators for power networks," in *Innovative Smart Grid Technologies Conference (ISGT), 2014 IEEE PES*. Piscataway, NJ, USA: IEEE Press, Feb 2014, pp. 1–5.
37. A. Baiocco, S. Wolthusen, C. Foglietta, and S. Panzieri, "A model for robust distributed hierarchical electric power grid state estimation," in *Innovative Smart Grid Technologies Conference (ISGT), 2014 IEEE PES*. Piscataway, NJ, USA: IEEE Press, Feb 2014, pp. 1–5.
38. F. D. Garcia and B. Jacobs, "Privacy-friendly energy-metering via homomorphic encryption," in *Proceedings of the 6th International Conference on Security and Trust Management*, ser. STM'10. Berlin, Heidelberg: Springer-Verlag, 2011, pp. 226–238.
39. K. Kursawe, G. Danezis, and M. Kohlweiss, "Privacy-friendly aggregation for the smart-grid," in *Proceedings of the 11th International Conference on Privacy Enhancing Technologies*, ser. PETS'11. Berlin, Heidelberg: Springer-Verlag, 2011, pp. 175–191. [Online]. Available: http://dl.acm.org/citation.cfm?id=2032162.2032172
40. N. Alamatsaz, A. Boustani, M. Jadliwala, and V. Namboodiri, "Agsec: Secure and efficient cdma-based aggregation for smart metering systems," in *Consumer Communications and Networking Conference (CCNC), 2014 IEEE 11th*, Jan 2014, pp. 489–494.
41. D. Kifer and A. Machanavajjhala, "No free lunch in data privacy," in *Proc. of the 2011 ACM SIGMOD Int.Conf.on Management of Data*, ser. SIGMOD '11. New York, NY, USA: ACM, 2011, pp. 193–204.

Index

A. V. D. M. Kayem et al. (eds.), *Smart Micro-Grid Systems Security and Privacy*,
Advances in Information Security 71, https://doi.org/10.1007/978-3-319-91427-5

Printed in the United States
By Bookmasters